GCSE
Success

Maths

Higher Tier

Exam Practice Workbook

Deborah Dobson, Phil Duxbury, Mike Fawcett and Aftab Ilahi

Number

Algebra

Ratio, Proportion and Rates of Change

Geometry and Measures

Probability

Statistics

Practice Exam

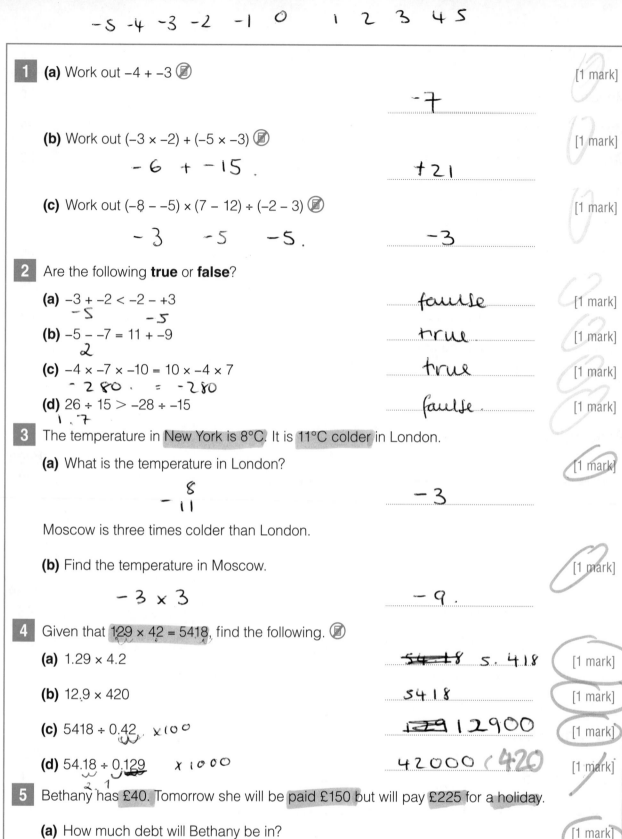

-5 -4 -3 -2 -1 0 1 2 3 4 5

1 **(a)** Work out −4 + −3 📱 [1 mark]

-7

(b) Work out (−3 × −2) + (−5 × −3) 📱 [1 mark]

- 6 + - 15.

+21

(c) Work out (−8 − −5) × (7 − 12) ÷ (−2 − 3) 📱 [1 mark]

- 3   - 5   - 5.

-3

2 Are the following **true** or **false**?

(a) −3 + −2 < −2 − +3 [1 mark]
-5   -5

faulse

(b) −5 − −7 = 11 + −9 [1 mark]
2

true

(c) −4 × −7 × −10 = 10 × −4 × 7 [1 mark]
- 280 . = -280

true

(d) 26 ÷ 15 > −28 ÷ −15 [1 mark]
1 . 7

faulse

3 The temperature in New York is 8°C. It is 11°C colder in London.

(a) What is the temperature in London? [1 mark]

- 8
- 11

−3

Moscow is three times colder than London.

(b) Find the temperature in Moscow. [1 mark]

- 3 × 3

- 9.

4 Given that 129 × 42 = 5418, find the following. 📱

(a) 1.29 × 4.2 [1 mark]

5418 5.418

(b) 12.9 × 420 [1 mark]

5418

(c) 5418 ÷ 0.42 ×100 [1 mark]

129 12900

(d) 54.18 ÷ 0.129 × 1000 [1 mark]
2 1

42000 (420

5 Bethany has £40. Tomorrow she will be paid £150 but will pay £225 for a holiday.

(a) How much debt will Bethany be in? [1 mark]

40 +150 = 190. 225
                - 190
                ___
                 35

£35

Bethany's father gives her $\frac{2}{5}$ of the amount she is in debt.

(b) How much debt is Bethany in now? 35 - 14 = 21 [2 marks]

£14

35 × 5 =
35 ÷ 5 = 7 × 2 = 14

Score 13/16

For more help on this topic, see Letts GCSE Maths Higher Revision Guide pages 4–5.

handwritten top: 3, 16   4, 12.
1, 48   6, 8
2, 24

1 Write down all the factors of 48 which are

(a) odd numbers [1 mark]

3, 1

(b) square numbers [1 mark]

16, 2, 4, 8

(c) multiples of 6 [1 mark]

24, 12, 6, 48

left margin: 3 6 9 12 15 18 21 24 27 30 33 36

2 James is thinking of a number between 1 and 50. It is 4 more than a prime number and 4 less than a square number. It is also a multiple of 3.

What could James's number be? [2 marks]

.... or ....

3 (a) Find the lowest common multiple (LCM) of 5 and 11. [1 mark]

55

(b) Find the highest common factor (HCF) of 14 and 35. [1 mark]

490   7

4 (a) Write 72 as a product of its prime factors. Give your answer in index form. [3 marks]

72 → 2, 36 → 12 → 6 → 3, 2, 3 → 2

$2^3 \times 3^2$

(b) Write 90 as a product of its prime factors. Give your answer in index form. [3 marks]

90 → 5, 9, 10 → 2, 3, 3

$3^2 * 5 * 2.$

(c) Find the highest common factor (HCF) of 72 and 90. [2 marks]

18

$2 \times 3 \times 3.$

(d) Find the lowest common multiple (LCM) of 72 and 90. [2 marks]

360

5 Bus A leaves the bus station every 12 minutes.
Bus B leaves the bus station every 28 minutes.
Bus A and Bus B both leave the bus station at 8.00am.

A = 8:12.
8:00 — B = 8:28.

When will they next both leave the bus station at the same time? [3 marks]

....

Score 14/20

For more help on this topic, see Letts GCSE Maths Higher Revision Guide pages 6–7.

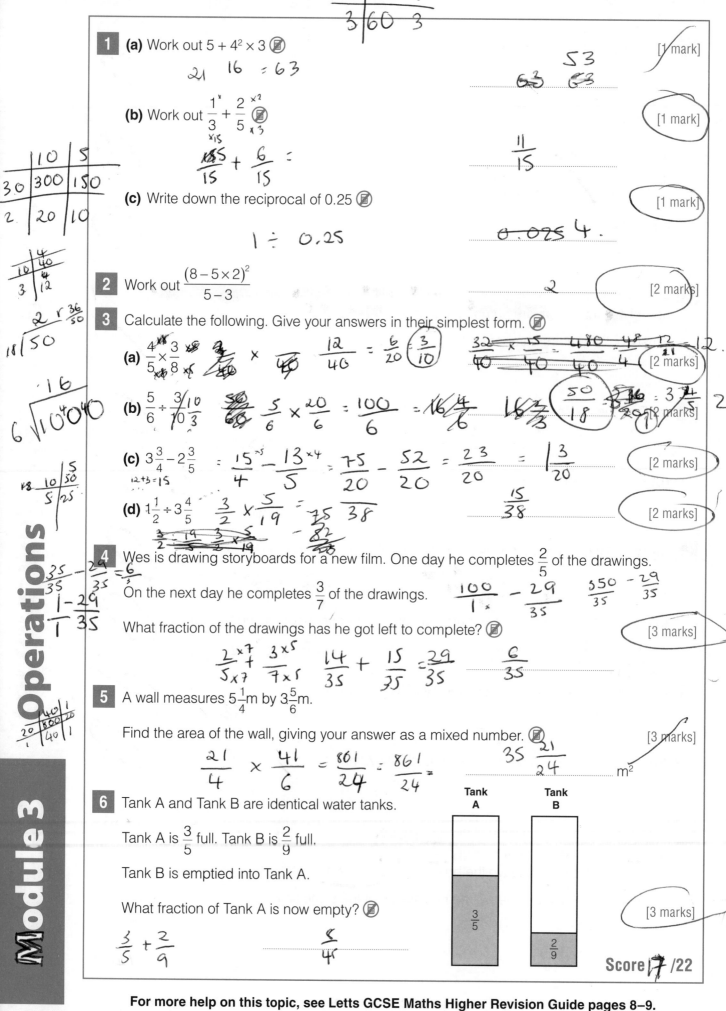

1 **(a)** Work out $5 + 4^2 \times 3$ 　[1 mark]

21　16 $= 63$

$\frac{20\ 1}{3\overline{)60\ 3}}$

53
63　63

(b) Work out $\frac{1}{3} + \frac{2}{5}$ 　[1 mark]

$\frac{15}{15} + \frac{6}{15} =$

$\frac{11}{15}$

(c) Write down the reciprocal of 0.25 　[1 mark]

$1 \div 0.25$

0.025　4.

2 Work out $\dfrac{(8 - 5 \times 2)^2}{5 - 3}$ 　2　[2 marks]

3 Calculate the following. Give your answers in their simplest form.

(a) $\frac{4}{5} \times \frac{3}{8}$ 　$\times \frac{12}{46} = \frac{6}{20}\left(\frac{3}{10}\right)$ 　$\frac{32}{40} \times \frac{15}{40} = \frac{480}{40} \frac{48}{4} \frac{12}{4} = 12$ 　[2 marks]

(b) $\frac{5}{6} \div \frac{3}{10}$ 　$\frac{5}{6} \times \frac{20}{6} = \frac{100}{6} = 16\frac{4}{6}$ 　$16\frac{2}{3}$ 　$\frac{50}{18}$ 　$\frac{16}{20} = 3\frac{4}{5}$ 　2　[2 marks]

(c) $3\frac{3}{4} - 2\frac{3}{5}$ 　$= \frac{15}{4} - \frac{13}{5} = \frac{75}{20} - \frac{52}{20} = \frac{23}{20} = 1\frac{3}{20}$ 　[2 marks]

(d) $1\frac{1}{2} \div 3\frac{4}{5}$ 　$\frac{3}{2} \times \frac{5}{19} = \frac{75}{38}$ 　$\frac{15}{38}$ 　[2 marks]

4 Wes is drawing storyboards for a new film. One day he completes $\frac{2}{5}$ of the drawings.

On the next day he completes $\frac{3}{7}$ of the drawings. 　$\frac{100}{1} - \frac{29}{35}$ 　$\frac{350}{35} - \frac{29}{35}$

What fraction of the drawings has he got left to complete? 　[3 marks]

$\frac{2}{5} + \frac{3}{7}$ 　$\frac{14}{35} + \frac{15}{35} = \frac{29}{35}$ 　$\frac{6}{35}$

5 A wall measures $5\frac{1}{4}$m by $3\frac{5}{6}$m.

Find the area of the wall, giving your answer as a mixed number. 　[3 marks]

$\frac{21}{4} \times \frac{41}{6} = \frac{861}{24} = \frac{861}{24} =$ 　$35\frac{21}{24}$ 　m²

6 Tank A and Tank B are identical water tanks.

Tank A is $\frac{3}{5}$ full. Tank B is $\frac{2}{9}$ full.

Tank B is emptied into Tank A.

What fraction of Tank A is now empty? 　[3 marks]

$\frac{3}{5} + \frac{2}{9}$ 　$\frac{5}{45}$

Score 17 /22

For more help on this topic, see Letts GCSE Maths Higher Revision Guide pages 8–9.

1 Write the following in order of value, smallest first.

$2^5 = 32$ $\sqrt[3]{64}$ 4 3^3 27 5^2 25 $\sqrt[4]{625}$ 5

[1 mark]

$4 = 5 \quad \sqrt[3]{64}, \quad \sqrt[4]{625}, \quad 5^2, \quad 3^3, \quad 2^5$

2 Find the value of x in each of the following.

(a) $2^x = 128$ 64   [1 mark]

(b) $\sqrt[3]{x} = 5$ 125   125.   [1 mark]

(c) $x^4 = 81$ 3   [1 mark]

3 (a) Write $4^6 \div 4^{-4}$ as a power of 4 in its simplest form.   [1 mark]

$4^{10}.$

(b) Write $\sqrt{32}$ as a surd in its simplest form.   [1 mark]

$\sqrt{4 \times 8}$ $4\sqrt{2}$

(c) Write 3^{-2} as a fraction in its simplest form.    [1 mark]

$\dfrac{16}{6}$

$2\dfrac{7}{9}$

4 (a) Write the following numbers in standard form.

(i) 321 000, $321 \times 10^{+5}$ [1 mark]

(ii) 0.000 605. 605×10^{-4} [1 mark]

(b) Calculate the following. Leave your answer in standard form.

(i) $(5 \times 10^3) \times (6 \times 10^{-8})$ [2 marks]

(ii) $(4.1 \times 10^4) + (3.4 \times 10^3)$ [2 marks]

5 Simplify the following. Leave your answer as a surd in its simplest form.

(a) $\sqrt{3}(4 - \sqrt{3})$ $4\sqrt{3} - \sqrt{9}$ [1 mark]

(b) $(2 + 3\sqrt{5})(4 - 2\sqrt{5})$ [2 marks]

6 Rationalise the denominator of the following surds.

(a) $\dfrac{4}{\sqrt{7}}$ $\dfrac{4 \times \sqrt{7}}{\sqrt{7} \times \sqrt{7}} = \dfrac{4\sqrt{7}}{7}$ [1 mark]

(b) $\dfrac{5}{2 - \sqrt{3}}$ [3 marks]

$\dfrac{5(2 - \sqrt{3})}{2 - \sqrt{3} \times 2 - \sqrt{3}} = \dfrac{5(2 - \sqrt{3})}{2 - 3}$

Score /20

Powers and Roots

Module 4

For more help on this topic, see Letts GCSE Maths Higher Revision Guide pages 10–11.

1 Write the following in order of value, smallest first. 📝

$\frac{12}{25}$ ~~$\frac{5}{4}$~~ $\frac{3}{8}$ 0.475 $\frac{17}{51}$

[1 mark]

....

2 Write 47.5% as a fraction in its simplest form.

[1 mark]

....

3 Which of the fractions $\frac{5}{8}$ and $\frac{7}{11}$ is closest to $\frac{2}{3}$? Show your working. 📝

[3 marks]

....

....

4 Calculate the following. 📝

(a) 3.7×4.9

[2 marks]

....

(b) $369.6 \div 1.4$

[2 marks]

....

5 Carl orders 34 Creamy Crisp Doughnuts for a church event.

Each doughnut costs £1.29

Find the total cost of the doughnuts. Show your working. 📝

[3 marks]

....

....

6 Write $0.\dot{4}$ as a fraction in its simplest form.

[1 mark]

....

7 Prove that $0.\dot{4}\dot{5}$ is equivalent to $\frac{5}{11}$

[2 marks]

....

8 Write $0.5\dot{3}\dot{6}$ as a fraction in its simplest form.

[2 marks]

....

Score   /17

For more help on this topic, see Letts GCSE Maths Higher Revision Guide pages 12–13.

1 Evaluate the following numbers to the degree of accuracy shown.

(a) $\sqrt{90}$ (2 decimal places)   9.49   [1 mark]

(b) 2.56^2 (3 decimal places)   6.554   [1 mark]

(c) $\sqrt[3]{175.6}$ (2 decimal places)   5.6   [1 mark]

2 **(a)** Round 404 928 to 2 significant figures.   [1 mark]

400000

(b) Round 0.04965 to 3 significant figures.   [1 mark]

0.0497

(c) Round π to 4 significant figures.   [1 mark]

3.1942

3 Violet runs a coffee lounge. She wants to make 35 cups of coffee from a 500g bag of coffee.

How much coffee should she use for each cup? Give your answer to 3 significant figures.

0.07×35 $\dfrac{35}{500}$ $500 \div 35$ 14.29   2.45. g   [2 marks]

4 Rachel works in a school office. She sends, on average, 72 letters home each week. Each letter costs 59p to send. There are 39 weeks in the school year.

Estimate the cost of sending the letters home for the whole year.    [3 marks]

$72 \times 39 = 2808. \times 059.$

$70 \times 40 \times 59$ £ 9 7746.48

5 Ethan is 1.73m tall to the nearest centimetre.

Write down the upper and lower bounds for his height.

$\dfrac{1.73}{2} = 0.865$ [2 marks]

Upper bound: ~~0.63~~ 2.595 m   Lower bound: 0.865 m

6 A water bottle holds 500ml of water to the nearest millilitre.

What is the maximum amount of water in a pack of 8 bottles?   [2 marks]

500×8 4000 ml

7 The volume of a shape can be found using the formula $v = \dfrac{m}{d}$, where m is mass and d is density.

If $m = 6.8$ to 2 significant figures and $d = 0.34$ to 2 significant figures, find the minimum and maximum value for v to 2 decimal places.   20   [4 marks]

Minimum value: ~~£~~ 10   Maximum value: 30

$\dfrac{20}{2} = 10$

Score 8 /19

For more help on this topic, see Letts GCSE Maths Higher Revision Guide pages 14–15.

Approximations   **Module 6**

Approximations

Module 6

1 Calculate the following. Write down all the figures on your calculator display.

(a) 4.5^4 .... [1 mark]

(b) $\sqrt[4]{2401}$ .... [1 mark]

(c) $\sqrt[7]{78125}$ .... [1 mark]

2 Calculate $\dfrac{4+3.9^2}{5.2\times 6.3}$

Write down all the figures on your calculator display. [2 marks]

....

3 Work out $(4.56 \times 10^{-2}) \div (3.8 \times 10^7)$, giving your answer in standard form. [1 mark]

....

4 Calculate the following using the fraction button on your calculator. Give your answer as a mixed number.

(a) $14\frac{2}{3}-11\frac{5}{7}$ .... [1 mark]

(b) $4\frac{3}{10}\times 7\frac{2}{9}$ .... [1 mark]

5 The Sun is 400 times further away from the Earth than the Moon.

The Sun is 1.496×10^8 km away from the Earth.

(a) How far away is the Moon from the Earth? Write your answer in standard form. [3 marks]

.... km

The Moon has a diameter of 3.48×10^3 km. The Sun has a diameter of 1.392×10^6 km.

(b) How many times bigger is the Sun's diameter compared with the Moon's diameter? Write your answer in standard form. [3 marks]

....

(c) Explain why the Sun and the Moon appear the same size in the sky. [1 mark]

....

....

(d) Find the distance between the Sun and the Moon when the Earth is directly between them. Write your answer in standard form to 2 significant figures. [3 marks]

.... km

Score /18

For more help on this topic, see Letts GCSE Maths Higher Revision Guide pages 16–17.

1 Simplify the following expressions. ✐

(a) $8a + 2 - 3a$ .... [1 mark]

(b) $6h + 3k + 9h - k$ .... [1 mark]

(c) $5a + 2b + 3c - 4a - 3a - 3c$ .... [1 mark]

(d) $3x^2 - 2x + 5x^2 + x + 1$ .... [1 mark]

2 Simplify the following expressions. ✐

(a) $q^2 p \div pq$ .... [1 mark]

(b) $\left(3p^4\right)^3$ .... [1 mark]

(c) $18f^7 g^2 \div 3f^2 g$ .... [1 mark]

(d) $\left(2p^2\right)^{-3}$ .... [1 mark]

3 Simplify the following expressions where possible. ✐

(a) $60k^2 \div 5$ .... [1 mark]

(b) $\sqrt[4]{a^{12}}$ .... [1 mark]

(c) $\sqrt[3]{p^6} \times \sqrt[6]{p^3}$ .... [1 mark]

(d) $p^2 \times p^3 \times \sqrt{p^3}$ .... [1 mark]

4 Evaluate ✐

(a) 4^{-3} .... [1 mark]

(b) $\left(\dfrac{2}{5}\right)^{-2}$ .... [1 mark]

(c) $32^{\frac{2}{5}}$ .... [1 mark]

(d) 16^0 .... [1 mark]

5 Which of the following are **equations** and which are **identities**? ✐

(a) $3x - 4 = (3x + 1) - (x - 5)$ .... [1 mark]

(b) $\dfrac{p+q}{pq} = \dfrac{1}{p} + \dfrac{1}{q}$ .... [1 mark]

(c) $x^2 + x^2 = 2x^2$ .... [1 mark]

(d) $x^2 + x^2 = x^4$ .... [1 mark]

(e) $\sqrt{a + b} = \sqrt{a} + \sqrt{b}$ .... [1 mark]

Score   /21

For more help on this topic, see **Letts GCSE Maths Higher Revision Guide pages 20–21.**

1 Expand and simplify $(2x-5y)(4x+3y)$ [2 marks]

2 Expand and simplify $p(2p+1)(2p-1)$ [2 marks]

3 Write $3(2x-1)+4(x+8)+5$ in the form $a(bx+c)$ where a, b and c are integers. [3 marks]

4 Factorise $a^2-19a+48$ [2 marks]

5 Simplify the expression $\dfrac{x^2-7x-18}{x^2-81}$ [3 marks]

6 Write $\dfrac{1}{a+2}-\dfrac{1}{a+3}$ as a single fraction. [2 marks]

7 Simplify $\dfrac{x+4}{x-2}\times\dfrac{2-x}{x+1}$ [2 marks]

8 Expand and simplify $(x+2)(x-3)(x+4)$ [4 marks]

9 Simplify the expression $\dfrac{x-9}{6}\div\dfrac{x^2-9x}{3}$ [4 marks]

10 Factorise $3x^2+19x-40$ [2 marks]

11 Expand and simplify $(2x+1)^3$ [3 marks]

12 Simplify the expression $\dfrac{x^2-5x-84}{x^2+5x-14}$ [3 marks]

Score /32

For more help on this topic, see Letts GCSE Maths Higher Revision Guide pages 22–23.

1 If $x = 4$, $y = -2$ and $z = \frac{1}{3}$, find the values of 📖

(a) $(x - y)^2$ [1 mark]

....

(b) $\frac{x^2}{z}$ [1 mark]

....

(c) $2z - y$ [1 mark]

....

(d) $xy + z$ [1 mark]

....

(e) $\sqrt{10x + 2y}$ [1 mark]

....

2 Rearrange $y = \frac{2 + 3x}{x - 8}$ to make x the subject. 📖 [3 marks]

....

3 Rearrange $s = ut + \frac{1}{2}at^2$ to make a the subject. 📖 [2 marks]

....

4 Rearrange $y = \frac{2}{x} + 5$ to make x the subject. 📖 [2 marks]

....

5 A suitcase weighs p grams when empty. When it is full of q grams of paper, it weighs r grams. Write down an equation for q in terms of p and r. 📖 [2 marks]

....

6 The formula to convert degrees Fahrenheit to degrees Celsius is $T_C = \frac{5(T_F - 32)}{9}$

(a) Use this formula to convert 88 degrees Fahrenheit (T_F) to degrees Celsius. [1 mark]

....

(b) Rearrange this formula to make T_F the subject. [2 marks]

....

(c) Hence convert 60 degrees Celsius to degrees Fahrenheit. [1 mark]

....

Score /18

For more help on this topic, see Letts GCSE Maths Higher Revision Guide pages 24–25.

Algebraic Formulae

Module 10

1 Solve the equation $5(x-2)-3(x-4)=4$ [2 marks]

2 Solve, by factorisation, the equation $x^2-3x=28$ [3 marks]

3 Solve, by factorisation, the equation $3x^2+x-10=0$ [3 marks]

4 $4x-3y=14$ and $x+y=7$, find the values of x and y. [2 marks]

$x =$ .... $y =$ ....

5 Solve the equation $x^2-8x-4=0$ by completing the square.
Give your answers in surd form. [3 marks]

6 Write $3x^2+24x+40$ in the form $a(x+b)^2+c$ [5 marks]

7 Solve the equation $3x^2-8x-13=0$ by using the quadratic formula.
Give your answers to 3 significant figures. [3 marks]

8 C is the curve with equation $y=x^2-x-2$ and L is the line with equation $y=2x+2$

If L and C intersect at two points A and B, find the exact length of AB. [5 marks]

9 Solve the simultaneous equations $y=8-x^2$ and $y=x-4$ [6 marks]

10 If you are given a graph of $y=2x^2-10x+7$, what straight line should you plot
in order to solve (graphically) the equation $4x^2-22x+9=0$? [4 marks]

11 $x^3-7x+2=0$ may be rearranged to give $x=\sqrt[3]{7x-2}$

By taking a starting value of $x_0=1$, use the iteration $x_{n+1}=\sqrt[3]{7x_n-2}$ to find a
solution to $x^3-7x+2=0$ to 2 decimal places. [3 marks]

Score /39

For more help on this topic, see Letts GCSE Maths Higher Revision Guide pages 26–27.

1 Solve the inequality $7x - 2 > 33$ 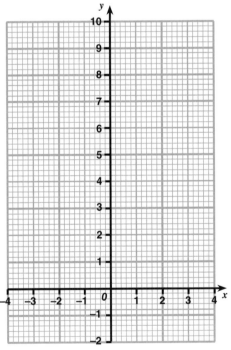   [2 marks]

....

2 State the inequalities as shown on the number lines below.   [1 mark]

(a)   ....   [1 mark]

(b)   ....   [1 mark]

(c)   ....   [1 mark]

3 List all integers x for which $x^2 < 10$ [2 marks]

....

4 Solve the inequality $x^2 - 9x + 8 \leqslant 0$, writing your answer using set notation.   [3 marks]

....

5 **(a)** On the grid below, draw sketches of $y = x^2$ and $y = x$ [2 marks]

(b) For what values of x is $x^2 < x$?   [1 mark]

....

(c) What are the solutions to $x^2 > x$ in set notation?   [2 marks]

....

6 Draw a rectangle with coordinates (2, 1), (5, 1), (2, 6) and (5, 6), joining the points with straight lines. Shade the rectangle.

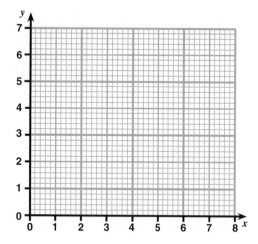

Describe the shaded area using inequalities. 🖩

[4 marks]

7 Shade the region corresponding to the following inequalities. 🖩

[4 marks]

$$y < x \qquad x + 3y < 15 \qquad y > 1$$

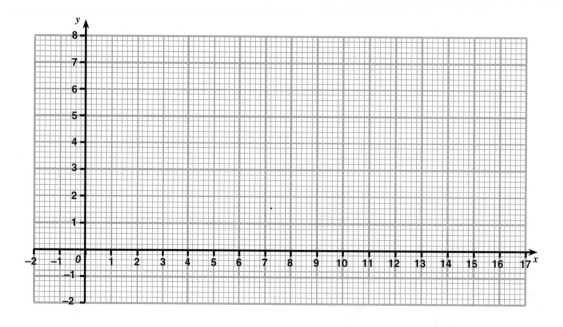

Score /23

For more help on this topic, see Letts GCSE Maths Higher Revision Guide pages 28–29.

1 Here are the first four terms of an arithmetic sequence.

41   32   23   14

Write down the next two terms in the sequence.   [2 marks]

....

2 Here are the first four terms of a quadratic sequence.

12   17   24   33

Write down the next two terms in the sequence.   [2 marks]

....

3 A term-to-term sequence is given by $U_{n+1} = 5U_n - 2$ where $U_1 = 1$

Write down the first five terms in this sequence.   [2 marks]

....

4 11, 15, 19, 23, … are the first four terms in an arithmetic sequence.

(a) Find an expression for U_n, the position-to-term formula.   [2 marks]

....

(b) Is 303 a number in this sequence? Explain your answer.   [2 marks]

....

5 7, 6, 7, 10, 15, … are the first five terms of a quadratic sequence.

(a) If the position-to-term sequence U_n is given by $U_n = an^2 + bn + c$, find the values of a, b and c.   [5 marks]

$a =$ .... $b =$ .... $c =$ ....

(b) Find U_{20}, the 20th term in the sequence.   [1 mark]

....

6 A geometric sequence has first term $a = 2\sqrt{3}$ and common ratio $r = 2\sqrt{3}$

(a) Write down an expression for the position-to-term formula, U_n [2 marks]

....

(b) Find an exact value for U_3, the third number in the sequence.   [1 mark]

....

Score   /19

For more help on this topic, see Letts GCSE Maths Higher Revision Guide pages 30–31.

Sequences

Module 13

1 Find the gradients of the following lines labelled A, B, C and D. 🖉 [4 marks]

A: $y = 6x \; 7$

B: $y = 8 \; 3x$

C: $2x + 3y + 4 = 0$

D: $y - x = 0$

Line A: Gradient is ....

Line B: Gradient is ....

Line C: Gradient is ....

Line D: Gradient is ....

2 From the following lines, write down a pair that is parallel and a pair that is perpendicular. 🖉 [2 marks]

A: $3y = x - 6$

B: $4x + 2y = 8$

C: $x = 3y - 12$

D: $2y - x + 2 = 0$

Lines .... and .... are parallel.

Lines .... and .... are perpendicular.

3 Write down the equation of the y-axis. [1 mark]

....

4 Write down the equation of the line parallel to $y = 5 - 6x$ that passes through the point (0, 9). [2 marks]

....

5 Work out the equation of the line parallel to $y = 5x - 2$ that passes through the point (−3, 1). [3 marks]

....

6 Find the equation of the line joining points (−4, −10) and (6, −5). [4 marks]

....

7 Find the equation of the line perpendicular to $3y = x + 15$ that passes through the point (5, −4). [5 marks]

....

Score /21

For more help on this topic, see Letts GCSE Maths Higher Revision Guide pages 32–33.

1 State the equation of the line of symmetry of the curve $y = 5x^2 - 30x + 12$ 📱 [2 marks]

....

2 Find (using algebra) the roots of the equation $x^2 - 7x - 44 = 0$ 📱 [3 marks]

....

3 Find (using algebra) the roots of the equation $3x^2 + 10x - 8 = 0$ 📱 [3 marks]

....

4 Consider the curve $y = x^2 + 6x + 4$ 📱

(a) By completing the square, find the coordinates of the minimum point on the curve. [3 marks]

....

(b) State the coordinates where the curve crosses the y-axis. [1 mark]

....

5 Consider the curve $y = (x - 2)(x - 4) - 48$ 📱

(a) Find where the curve crosses the x-axis. [4 marks]

....

(b) State the coordinates where the curve intercepts the y-axis. [1 mark]

....

(c) Find the equation of the line of symmetry. [1 mark]

....

6 (a) By completing the square, write $3x^2 - 12x + 22$ in the form $p(x + q)^2 + r$ 📱 [5 marks]

....

(b) Hence find the coordinates of the minimum point on the curve
$y = 3x^2 - 12x + 22$ 📱 [1 mark]

....

7 By completing the square, find the maximum point on the curve
$y = 25 + 10x - x^2$ 📱 [4 marks]

....

Score /28

For more help on this topic, see Letts GCSE Maths Higher Revision Guide pages 34–35.

Quadratic Functions

Module 15

1 Given $f(x) = 5x + 3$, find expressions for 🔲

   (a) $f^{-1}(x)$ [2 marks]

                                                            ....

   (b) $ff(x)$ [2 marks]

                                                            ....

2 Sketch the graphs of $y = \dfrac{1}{x}$ and $y = \dfrac{-4}{x}$ on the same axes. 🔲   [4 marks]

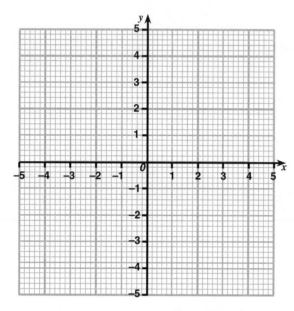

3 Given $f(x) = \dfrac{1}{x}$ and $g(x) = 4x - 6$, find 🔲

   (a) $g\left(\dfrac{1}{2}\right)$ [1 mark]

                                                            ....

   (b) $f\left(\dfrac{-1}{3}\right)$ [1 mark]

                                                            ....

   (c) $fg(1) - gf(1)$ [2 marks]

                                                            ....

4 On the grid on page 21, sketch graphs of 🔲

   (a) $y = 2^x$ [2 marks]

   (b) $y = 2^x + 3$ [2 marks]

   (c) $y = 2^{x+3}$ [2 marks]

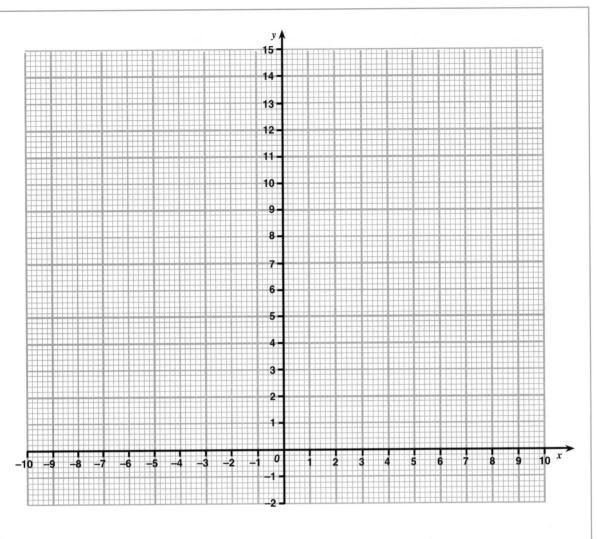

5 The graph of $y = \cos x$ transforms to $y = 2 + \cos(x - 90°)$ through a sequence of two transformations. Describe each transformation. 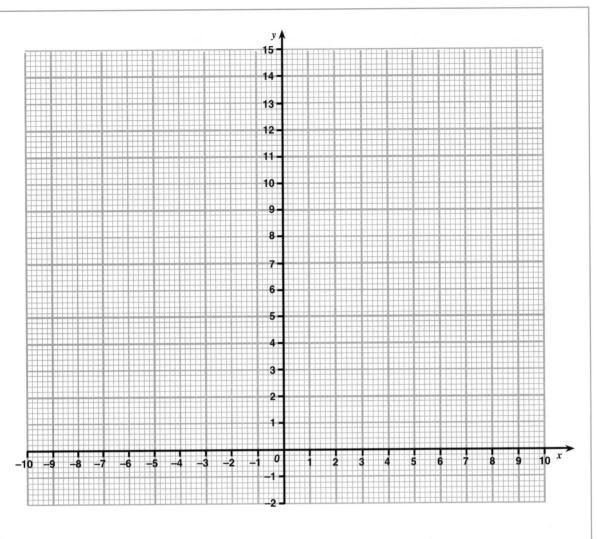 [4 marks]

....

....

....

6 Given $p(x) = \dfrac{3x+1}{1-x}$, find a simplified expression for $p^{-1}(x)$ [5 marks]

....

7 The graph of $y = x^2 - 2x + 5$ is found after applying two transformations to the graph of $y = x^2$

By writing $x^2 - 2x + 5$ in the form $(x - a)^2 + b$, describe the two transformations. [6 marks]

....

....

Score /33

For more help on this topic, see Letts GCSE Maths Higher Revision Guide pages 36–37.

1 Describe the shape defined by the equation $x^2 + y^2 = 2$ 🖩 [3 marks]

....

2 Consider the circle having equation $4x^2 + 4y^2 = 9$ 🖩

(a) Sketch the circle. [3 marks]

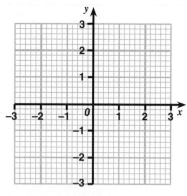

(b) Find an equation for the tangent to the circle at the point $\left(\dfrac{1}{2}, \sqrt{2}\right)$, giving your answer in the form $ax + by + c = 0$ [4 marks]

....

3 Consider the graph of $y = 12x - 4x^2$ shown.

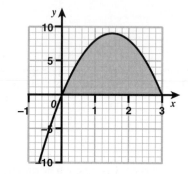

(a) Estimate the area bounded by the curve and the x-axis by splitting the area into six strips. [6 marks]

.... units^2

(b) Find an estimate for the equation of the tangent to the curve at the origin. [5 marks]

....

4 Given a circle $x^2 + y^2 = a^2$, prove that the equation of the tangent at the point (h, k) lying on the circumference is given by $hx + ky = a^2$ 🖩 [6 marks]

....

....

....

....

....

Score /27

For more help on this topic, see Letts GCSE Maths Higher Revision Guide pages 38–39.

1 Calculate the area of a rectangle with width 850mm and length 1.25m. [2 marks]

....

2 Convert 3.4km^2 to metres^2. 🖩 [2 marks]

....

3 If 1cm^3 of water has a mass of 1g, what is the mass of 1m^3 of water in kilograms? 🖩 [2 marks]

....

4 5 miles = 8km

Which is faster, 68mph or 110km/h? [2 marks]

....

5 Convert 24m/s to kilometres per hour. [3 marks]

....

6 Shahida works 30 hours each week and earns £18.40/h.

Bob works 35 hours each week and earns £626.00.

Leila works 32 hours each week and earns £30 000 per annum.

(a) Who has the best hourly rate? [3 marks]

....

(b) Who earns the most per year? [2 marks]

....

7 Here are two pieces of gold:

Block A has a volume of 24cm^3 and a mass of 463g.

The volume of block B is 30cm^3 and it contains exactly the same quality of gold.

A   B

(a) Calculate the density of gold. [2 marks]

....

(b) Calculate the mass of block B. [2 marks]

....

(c) If the price of gold is £23.50/g, calculate the total value of both pieces. [2 marks]

....

Score /22

For more help on this topic, see Letts GCSE Maths Higher Revision Guide pages 42–43.

Module 18

1 A model plane is made using a scale of 1:20 🔲

(a) If the wingspan of the model is 3m, what is the wingspan of the actual plane? [1 mark]

....

(b) If the plane is 57m long, how long is the model? [2 marks]

....

2 Here is a map with a scale of 1:5000000 🔲

(a) What is the distance on the map between Bristol and Leeds? [1 mark]

....

(b) What is the actual distance? [2 marks]

....

(c) Measure the bearing of Exeter from Leeds. .... [2 marks]

(d) What is the bearing of Leeds from Exeter? .... [2 marks]

3 A cruise ship sets off from harbour A for 100km on a bearing of 075° to point B.
It then changes to a bearing of 220° and sails at 15km/h for 6 hours to point C.
Finally it heads straight back to port.

(a) Draw an accurate diagram to show the journey using a scale of 1cm : 20km   [2 marks]

(b) Measure the bearing and distance of the last leg of the journey from C to A.   [2 marks]

Bearing: ....   Distance: .... km

4 A map is drawn to a scale of 1 : 25 000

Villages S and R are 18cm apart on the map.

(a) Work out the actual distance between S and R in kilometres.   [2 marks]

....

R is due West of S. T is another village. The bearing of T from R is 040° and
the bearing of T from S is 320°.

(b) Draw a diagram to show the positions of S, R and T.   [2 marks]

Score   /18

For more help on this topic, see Letts GCSE Maths Higher Revision Guide pages 44–45.

Scales, Diagrams and Maps

Module 19

1 Circle the numbers which are **not** equivalent to 3.75 ▱ [2 marks]

$\frac{30}{8}$ 37.5%   375% $\frac{14}{5}$ $3\frac{5}{8}$ $\frac{750}{200}$

2 Write the following numbers in ascending order. ▱ [2 marks]

0.21   20% $\frac{3}{10}$ 0.211 $\frac{2}{9}$

3 Is 2.125 or $2\frac{4}{5}$ closer to $2\frac{1}{2}$? Explain your reasoning. ▱ [2 marks]

4 On a new estate of 32 houses, $\frac{3}{8}$ have two bedrooms. $\frac{5}{6}$ of the two-bedroom houses have a garage.

What percentage of the whole estate is represented by two-bedroom houses with a garage? [2 marks]

5 Claire makes soft toys to sell at a Christmas market.

(a) Each dog toy costs £3.45 to make and Claire sells them for £5.99.

What is her percentage profit? [2 marks]

(b) A tiger toy costs 15% more to make than a dog toy and she makes 80 tiger toys. Claire sells 55 of them for £6.99 and the rest at the reduced price of £4.

What percentage profit does she make on tiger toys? [4 marks]

6 At Mathstown School 55% of the students are girls. 40% of the girls and 65% of the boys have school lunch.

(a) What percentage of students at the school have school lunch? [3 marks]

(b) What fraction of the boys do not have school lunch? [2 marks]

Score   /19

For more help on this topic, see Letts GCSE Maths Higher Revision Guide pages 46–47.

1 Simplify these ratios and circle the odd one out. You must show all your working. 🔲 [2 marks]

£4:£6   10:15   20cm:3m   750g:1.125kg   40 seconds:1 minute

....

....

....

2 Jane is making 'mist blue' paint for her room. She mixes navy blue, grey and white paint in the ratio 1:2:7

(a) How much of each colour does Jane need to make 2 litres of paint?   [3 marks]

Navy blue: .... ml   Grey: .... ml   White: .... ml

(b) Jane finds she has $\frac{3}{4}$ litres of navy blue, 1200ml of grey and 6 litres of white paint.

What is the maximum amount of 'mist blue' she can make?   [2 marks]

.... litres

3 The ratio of A:B is 5:8

Complete this statement. A is $\frac{\square}{\square}$ of B. 🔲   [1 mark]

4 This is a recipe for shortbread:

Makes 15 biscuits			
110g butter	50g sugar	175g flour	50g chocolate chips

(a) Amil has 70g of sugar. How many biscuits can he make?   [2 marks]

....

(b) How much flour is needed to make 12 biscuits?   [2 marks]

.... g

5 Lucy makes green paint by mixing yellow and blue paint in the ratio 5:2
Blue paint costs £30 for 5 litres and yellow paint costs £28 for 7 litres.

Lucy sells her paint for £4.50 per litre. Will she make a profit? Show your working to justify your decision. 🔲   [3 marks]

....

....

Score /15

Ratio

Module 21

For more help on this topic, see Letts GCSE Maths Higher Revision Guide pages 48–49.

1 If Shabir has 250ml of soup for her lunch, how many kilocalories of energy will she get?

.... [2 marks]

2 Leon changes £500 to euros at the rate shown and goes to France on holiday.

£1 = 1.29 euros	£1 = 187.99 Japanese yen
£1 = 1.56 US dollars	£1 = 97.04 Indian rupees

(a) How many euros does he take on holiday? .... [1 mark]

Leon spends €570.

(b) He changes his remaining euros on the ferry where the exchange rate is £1 : €1.33

How much in pounds sterling does he take home? .... [2 marks]

3 James' dairy herd of 80 cattle produces 1360 litres of milk per day.

(a) If James buys another 25 cattle and is paid 30p/litre, what will his annual milk income be? .... [4 marks]

(b) If 6 tonnes of hay will last 80 cattle for 10 days, how long will the same amount of hay last the increased herd? [2 marks]

....

4 Triangles *PQR* and *STU* are similar.

Find the missing lengths *PR* and *TU*.

PR = ...

TU = ...

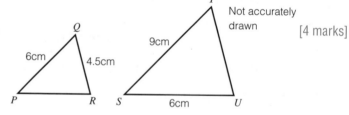

Not accurately drawn   [4 marks]

5 Two similar cylinders P and Q have surface areas of 120cm² and 270cm².

If the volume of Q is 2700cm³, what is the volume of P? [3 marks]

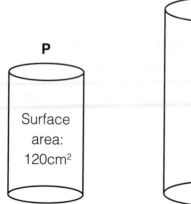

.... cm³

Score   /18

For more help on this topic, see **Letts GCSE Maths Higher Revision Guide** pages 50–51.

1 (a) Peter invests £10000 in a savings account which pays 2% compound interest per annum.

How much will his investment be worth after four years? [2 marks]

....

(b) Paul invests £10000 in company shares.
In the first year the shares increase in value by 15%.
In the second year they increase by 6%.
In the third year they lose 18% of their value.
In the fourth year the shares increase by 1%.

What is his investment worth after four years? .... [3 marks]

2 Lazya invests £6500 at 3% compound interest for three years. She works out the first year's interest to be £195. She tells her family she will earn £585 over three years.

Is she right? Show working to justify your decision. [3 marks]

....

....

3 This graph shows a tank being filled with water.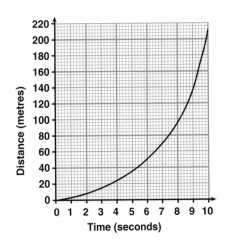

(a) How deep is the water when the tank is full?

.... [1 mark]

(b) Between what times is the tank filling fastest?

.... [1 mark]

(c) Work out the rate of decrease of water level as the tank empties. .... [1 mark]

4 This graph shows the distance travelled by a cyclist for the first 10 seconds of a race.

(a) Work out the cyclist's average speed for the first 10 seconds.

.... [2 marks]

(b) Estimate the actual speed at 5 seconds.

.... [3 marks]

Score /16

For more help on this topic, see Letts GCSE Maths Higher Revision Guide pages 52–53.

1 Using only a ruler and a pair of compasses, construct a triangle with sides 8cm, 7cm and 9cm.

[2 marks]

2 Using only a ruler and a pair of compasses, construct a rectangle with one side 8cm and an area of 28cm².

[2 marks]

3 Using only a ruler and a pair of compasses, draw a line *PQ* such that angle *PQR* is 90°.

[2 marks]

Q _____ *R*

4 This is a plan of a field using a scale of 1cm : 10m

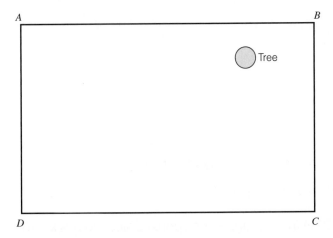

The farmer wants to install a water trough. It must be

- at least 15m from the tree
- more than 20m from the fence DC
- nearer to AB than to AD.

Shade the area where the farmer can put the trough. [3 marks]

5 Using only a ruler and a pair of compasses, construct an angle of 30°. [3 marks]

Score   /12

For more help on this topic, see Letts GCSE Maths Higher Revision Guide pages 56–57.

Constructions

Module 24

1 Explain, using a diagram if required, why a regular pentagon will not tessellate. [2 marks]

....

....

....

....

2 The internal angles of a regular polygon are 157.5°

How many sides has the polygon? .... [2 marks]

3 ABIJKL and BCDEFGHI are regular polygons. Calculate angle JIH. [3 marks]

....

4 Calculate angle FED. State all your reasons. [3 marks]

....

....

....

....

5 Lines BC and DE are parallel. Calculate angle CPR, giving all your reasons. [2 marks]

....

....

....

Score /12

For more help on this topic, see Letts GCSE Maths Higher Revision Guide pages 58–59.

1 Draw lines to match each triangle with its correct description. [3 marks]

A   **B**   **C**   **D**   **E**

| Scalene | Right-angled isosceles | Equilateral | Obtuse scalene | Obtuse isosceles |

2 Use this diagram to prove the angle sum of a triangle is 180°. [3 marks]

P ———— B ————→ Q

b

a c

R A C S

....
....
....
....

3 Use this diagram to find the angle sum of an octagon. [3 marks]

....

4 Complete this sentence with the name of the correct quadrilateral.

A .... has two pairs of equal sides but only one pair of equal angles. [1 mark]

5 Find angle x, giving all your reasons. [3 marks]

x

....
....
....
....

Score /13

For more help on this topic, see Letts GCSE Maths Higher Revision Guide pages 60–61.

Module 26

1 Circle the two similar triangles. [1 mark]

A 4 3 5.5

B 3 2 4.5

C 6.875 4.5 3

D 6 3.5 4.5

E 6¾ 3 4.5

2 Circle the two congruent triangles. [1 mark]

G 4 40° 5

H 4 5 40°

K 8 40° 10

I 4 87° 5 53°

J 4 53° 5

3 *PQRS* is a parallelogram. Prove that triangle *PQS* is congruent to triangle *RSQ*. [3 marks]

Q   R

P   S

....

....

....

....

4 State whether these triangles are congruent and give your reasons. [2 marks]

X 3cm Y 4cm Z

K 4cm L 5cm M

....

....

....

5 Cylinder A and cylinder B are mathematically similar.

The volume of cylinder A is 320cm³. What is the volume of B? [4 marks]

.... cm³

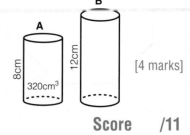

A 8cm 320cm³

B 12cm

Score   /11

For more help on this topic, see Letts GCSE Maths Higher Revision Guide pages 62–63.

1 (a) Describe the transformation
   that moves A on to B.   [2 marks]

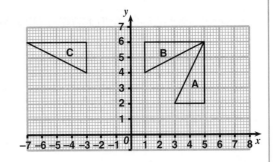

....

....

....

(b) Describe the transformation that moves B on to C.   [2 marks]

....

2 Describe fully the single transformation
   that maps P on to Q.   [3 marks]

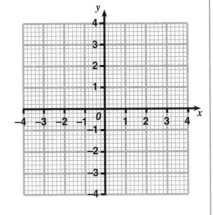

....

....

....

3 (a) Plot the triangle T with coordinates
   (3, 3), (3, 1) and (4, 1).   [1 mark]

(b) Rotate T 90° clockwise with centre
   (0, 0) and label the image V.   [2 marks]

(c) Reflect V in the *y*-axis and label
   the image W.   [2 marks]

(d) What single transformation maps T
   directly on to W?   [2 marks]

4 Enlarge this shape with

(a) scale factor 2, centre (3, 5)   [2 marks]

(b) scale factor 0.5, centre (0, 0)   [2 marks]

(c) scale factor –1, centre (0, 2).   [2 marks]

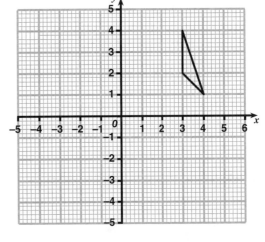

Score   /20

For more help on this topic, see Letts GCSE Maths Higher Revision Guide pages 64–65.

Transformations

Module 28

1 Complete the following statements with the correct word.

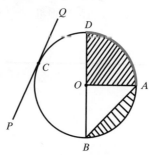

(a) *OA* is the _____ of the circle. [1 mark]

(b) *BD* is a _____ . [1 mark]

(c) *PCQ* is a _____ to the circle at *C*. [1 mark]

(d) The curved line around the full circle is the _____ . [1 mark]

(e) The blue line *AD* is an _____ . [1 mark]

(f) The shaded area *OAD* is a _____ . [1 mark]

(g) The shaded area between *A* and *B* is a _____ . [1 mark]

(h) *OAB* is an _____ triangle. [1 mark]

2 Explain, with reasons, whether each statement is **true** or **false**.

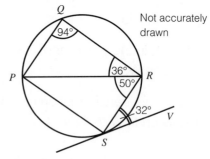

Not accurately drawn

(a) *PR* is a diameter of the circle. [1 mark]

....

....

(b) Angle *PSR* = 86° [1 mark]

....

....

(c) *SV* is a tangent to the circle. [2 marks]

....

....

3 PQRS are points on a circle centre O.

Work out the size of these angles, giving reasons for your answers.

(a) PRS [2 marks]

....

....

....

(b) PRQ [3 marks]

....

....

....

....

4 **(a)** Calculate these angles, giving reasons for your answers.

(i) ABD [2 marks]

....

....

....

....

(ii) BDA [2 marks]

....

....

(b) Prove that triangle ABD is isosceles. [2 marks]

....

....

(c) Prove that triangles CDF and ABF are similar. [2 marks]

....

....

....

Score /25

For more help on this topic, see Letts GCSE Maths Higher Revision Guide pages 66–67.

1 What 3D shape has 6 faces, 10 edges and 6 vertices?

[2 marks]

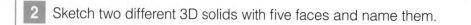

2 Sketch two different 3D solids with five faces and name them.

[3 marks]

3 Draw the plan and elevations of this 3D shape.

[4 marks]

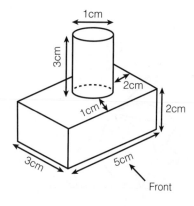

1cm

3cm

2cm

1cm

2cm

3cm

5cm

Front

4 Sketch the 3D shape shown by this plan and elevations.

[3 marks]

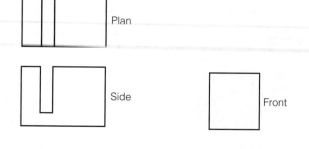

Plan

Side

Front

Score /12

For more help on this topic, see Letts GCSE Maths Higher Revision Guide pages 68–69.

1 Calculate the area and perimeter of these shapes.

(a) [3 marks]

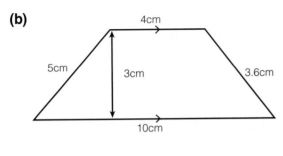

Area: .... m²   Perimeter: .... m

(b) [3 marks]

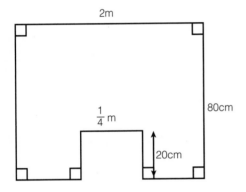

Area: .... cm²   Perimeter: .... cm

2 Calculate the area of this shape. [2 marks]

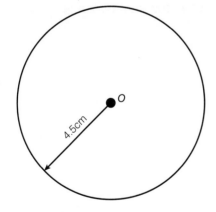

....

3 Calculate the circumference of this circle. Give your answer to 2 significant figures. [3 marks]

.... cm

4 Calculate the area of a circle with diameter 8cm. Give your answer in terms of π. [3 marks]

....

5 What is the perimeter of this shape? [2 marks]

8cm

4cm

5cm

.... cm

6 Calculate the surface area of a cylinder with radius 4cm and height 5cm.
Give your answer in terms of π. [3 marks]

.... cm²

7 This shape is a cone of vertical height 4cm sitting on a cube of side 3cm.
Calculate its volume. [3 marks]

4cm

3cm

....

8 A sphere of radius 3.5cm and a cube with sides *a* cm have the same volume.
Find *a*. Give your answer to 3 significant figures. [3 marks]

.... cm

9 Calculate the volume of a hemisphere with radius 5cm.
Give your answer in terms of π. [3 marks]

....

Score /28

For more help on this topic, see Letts GCSE Maths Higher Revision Guide pages 70–71.

1 **(a)** Work out the size of angle p.　　　　　　　　　　　[3 marks]

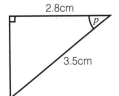

2.8cm

p

3.5cm

....

(b) Find the length KL. Give your answer to 3 significant figures.　　[3 marks]

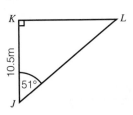

K　　　　　L

10.5m

51°

J

.... m

2 Calculate angle ABC.　　　　　　　　　　　　　　　　[3 marks]

A

8.1

63°

C　　6.3　　B

....

3 Find q, leaving your answer as a square root. 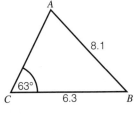　　[3 marks]

7cm

$q$cm

9cm

....

4 **(a)** Write down the value of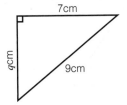

　　(i) sin 45°　　　　　　　　　.......................................　[1 mark]

　　(ii) tan 45°　　　　　　　　.......................................　[1 mark]

(b) Calculate a and b.　　　　　　　　　　　　　　[4 marks]

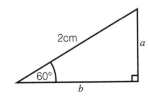

2cm

a

60°

b

　　$a = $ ....　$b = $ ....

5 Find the length of the largest pencil that can just fit in a box 14cm by 6cm by 3cm.　[2 marks]

.... cm

Score　/20

For more help on this topic, see Letts GCSE Maths Higher Revision Guide pages 72–73.

1 Here are base vectors **r**, **s** and **t** drawn on isometric paper.

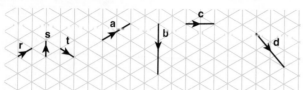

Write each of the vectors **a** to **d** in terms of **r**, **s** and **t**. [4 marks]

a = ...   **b** = ...   **c** = ...   **d** = ...

2 *ABCD* is a parallelogram.
BE is a straight line with *AB = AE*.
$\overrightarrow{AB}$ = **b**, $\overrightarrow{AD}$ = **d**.

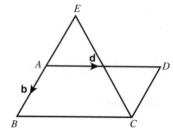

(a) Write $\overrightarrow{BE}$ in terms of **b** and **d**.   .... [1 mark]

(b) Write $\overrightarrow{AC}$ in terms of **b** and **d**.   .... [2 marks]

(c) Write $\overrightarrow{CE}$ in terms of **b** and **d**.   .... [2 marks]

3 *PQR* is an equilateral triangle.
OPQ is an isosceles triangle with *OQ = QR*

$\overrightarrow{OP}$ = **p**

$\overrightarrow{OQ}$ = **q**

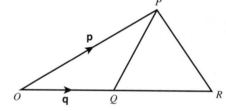

Find $\overrightarrow{OR}$, $\overrightarrow{PQ}$ and $\overrightarrow{PR}$. [3 marks]

$\overrightarrow{OR}$ = ... $\overrightarrow{PQ}$ = ... $\overrightarrow{PR}$ = ...

4 If *A* is the midpoint of *OP* and *B* is the midpoint of *OQ*, use vectors to
prove that *AB* is parallel to *PQ*. [3 marks]

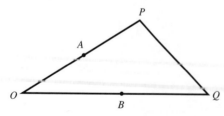

....

....

Score   **/15**

For more help on this topic, see Letts GCSE Maths Higher Revision Guide pages 74–75.

1 There are three bananas, five oranges, four apples and two mangoes in a fruit bowl. A piece of fruit is taken at random.

Find the probability that it is

(a) an apple .... [1 mark]

(b) a banana or a mango .... [1 mark]

(c) not an orange. .... [1 mark]

2 Paige wants to test whether she can predict the suit of a card drawn from an ordinary shuffled deck of playing cards. She does 120 tests.

How many correct predictions do you think she will make? [2 marks]

....

3 The table shows the probability that Mark will choose a certain activity when he gets home from school on any given day.

Activity	Watch TV	Do homework	Play on computer	Go for a bike ride	Other
Probability	0.2		0.3	0.25	0.1

Find the probability that Mark will do homework on any given day. [2 marks]

....

4 Brian bought a dice from the joke shop. He rolled it 25 times and recorded the results. Here are the results:

Number	1	2	3	4	5	6
Frequency	0	3	2	4	4	12

(a) Find the relative frequency of rolling a 6. [1 mark]

....

Brian rolls the dice another 250 times.

(b) How many times would he be expected to get an even number? [3 marks]

....

5 Morgan spins the spinner until she gets a 4.

Find the probability that she will stop after exactly two spins. [2 marks]

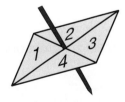

Score /13

For more help on this topic, see Letts GCSE Maths Higher Revision Guide pages 78–79.

1 Fifty students are voting for a new class president. They can each vote for a maximum of two people from Luke, Dan and Jack.

Luke gets 26 votes altogether
7 students vote for Dan and Luke
2 students vote for Dan and Jack

Dan gets 17 votes altogether
4 students vote for Luke and Jack
3 students don't vote

(a) Complete the Venn diagram to show this information. The universal set ℰ contains all 50 students.

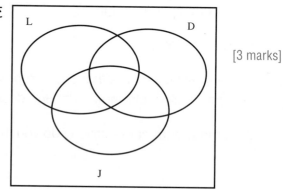

[3 marks]

(b) Find the probability that a student chosen at random votes for Jack only.   [1 mark]

....

(c) Find the probability that a student who voted for Luke also voted for Dan.   [2 marks]

....

2 Katy is a midwife. If she works on Christmas Day, then the probability that she will have to work on New Year's Day is 0.25

If she does not work on Christmas Day, then the probability that she will have to work on New Year's Day is 0.85

The probability that Katy will have to work on Christmas Day is 0.7

(a) Complete the tree diagram.   [2 marks]

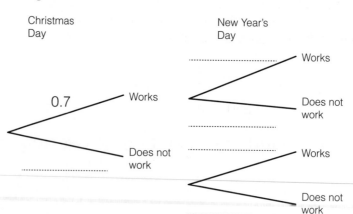

(b) Calculate the probability that Katy will only have to work on one of the two days.   [2 marks]

....

Score   /10

For more help on this topic, see Letts GCSE Maths Higher Revision Guide pages 80–81.

1 A washing powder manufacturer conducts a survey of the actual weight of powder in the 500g boxes on the production line. A sample of 12 boxes is taken over a period of 1 hour:

495, 503, 490, 505, 490, 498, 500, 510, 505, 498, 498, 501

(a) What is the mode for the sample? [1 mark]

....

(b) Find the median. [1 mark]

....

(c) The manufacturer produces 10 000 boxes per hour.
Give two reasons why this may not be a good sample. [2 marks]

....

....

2 The table shows the weights of 160 different species of birds in the forest.

Weight (g)	Frequency
$30 \leqslant w < 40$	20
$40 \leqslant w < 50$	25
$50 \leqslant w < 60$	22
$60 \leqslant w < 80$	44
$80 \leqslant w < 100$	32
$100 \leqslant w < 120$	15
$120 \leqslant w < 150$	2
	160

(a) Which class interval contains the median? [1 mark]

....

(b) Calculate an estimate of the mean weight of the birds. [3 marks]

....

(c) What percentage of the birds weigh less than 80g? [2 marks]

....

3 There are five cards with an integer on each. The mean of all the numbers on the cards is 7. Both the mode and the median of the numbers are 6.

What numbers could be on the cards? 📝 [3 marks]

....

....

Score /13

For more help on this topic, see Letts GCSE Maths Higher Revision Guide pages 84–85.

1 The table shows the sales figures for a car dealership over two different weeks.

Sales	Week 1	Week 2
Monday	10	25
Tuesday	25	20
Wednesday	29	15
Thursday	40	19
Friday	16	

(a) There were 100 car sales in total in week 2.
Complete the table.   [1 mark]

(b) Show this data in a suitable chart.   [3 marks]

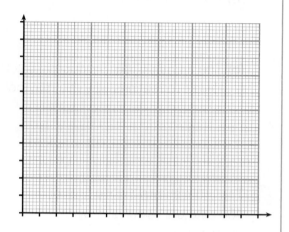

2 The manager of a pizza shop conducts a survey of favourite pizzas. She wants to show this information on a pie chart.

	Frequency	Angle
Vegetarian	45	60°
Seafood	80	
Meat	50	
Chicken	40	
Mushroom		

Complete the table.   [2 marks]

3 The table shows the sales of a particular type of mobile phone over an eight-year period.

Year	1	2	3	4	5	6	7	8
Sales (millions)	1.4	11.6	20.7	40	72.3	125	150.3	169.2

(a) By plotting a suitable graph, use this information to predict the sales for year 9.

....   [4 marks]

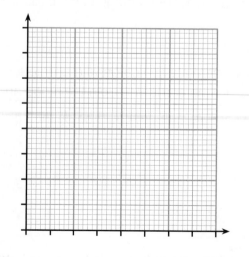

(b) Comment on the reliability of using your graph to predict sales in year 14. [2 marks]

....

....

4 The speeds of 12 drivers and their ages were recorded in the table below. John says that 'younger drivers drive too fast'.

Age (years)	20	32	24	30	22	40	35	34	42	22	38	32
Speed (mph)	38	30	37	32	39	30	32	33	28	36	27	34

(a) Use an appropriate diagram to comment on whether the data supports John's conclusion.

.... [3 marks]

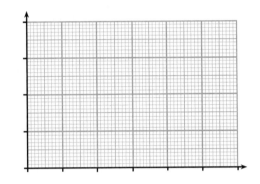

(b) Describe the relationship between age and driving speed.

.... [1 mark]

5 Wasim hypothesises that students travel further to get to college than they used to. In a survey in 2005, the mean weekly travelling distance to a college was 25.8km. The histogram below shows the results of another survey taken in 2015.

Weekly travelling distance

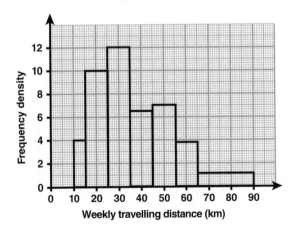

Does the evidence support the hypothesis? [5 marks]

....

....

....

Score   /21

For more help on this topic, see Letts GCSE Maths Higher Revision Guide pages 86–87.

1 The test scores of 24 students in a history exam are given below.

Class A	33	45	67	83	56	23	57	45	73	43	26	35
Class B	33	27	40	44	78	28	49	38	54	32	51	74

(a) Calculate the mean for each of the two classes. [2 marks]

....

....

(b) Draw box plots for both classes. [4 marks]

....

....

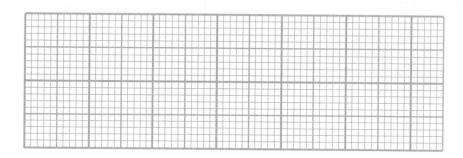

(c) Which class did better? Give reasons for your answer. [3 marks]

....

....

....

2 The table below shows the results of a survey on the cost of car insurance for 30–50-year-olds.

(a) Construct a cumulative frequency curve to represent the data. [3 marks]

Cost (£)	Frequency
$100 \leqslant c < 150$	10
$150 \leqslant c < 200$	38
$200 \leqslant c < 250$	48
$250 \leqslant c < 300$	31
$300 \leqslant c < 350$	20
$350 \leqslant c < 400$	8
$400 \leqslant c < 500$	5

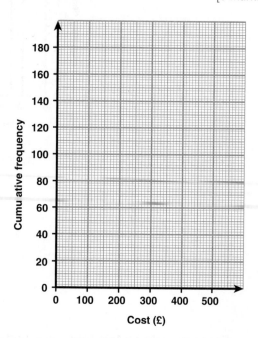

(b) Draw a box plot for this data. [2 marks]

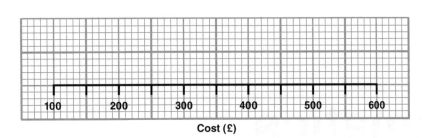

Cost (£)

(c) The same survey was conducted for drivers younger than 30 years old. The results are shown on this plot.

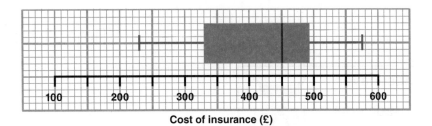

Cost of insurance (£)

Make two comparisons between the two sets of data. [2 marks]

Comparison 1: ....

....

Comparison 2: ....

....

(d) Use your cumulative frequency curve to estimate how many 30–50-year-olds pay more than £325 for car insurance. Explain why this is an estimate. [2 marks]

....

....

....

....

Score   /18

For more help on this topic, see Letts GCSE Maths Higher Revision Guide pages 88–89.

Module 38

GCSE
Mathematics
Higher tier

Paper 1   Time: 1 hour 30 minutes

For this paper you must have:

- mathematical instruments

You must **not** use a calculator.

Instructions

- Use black ink or black ball-point pen. Draw diagrams in pencil.
- Read each question carefully before you start to write your answer.
- Diagrams are not accurately drawn unless otherwise stated.
- Answer **all** the questions.
- Answer the questions in the space provided.
- In all calculations, show clearly how you work out your answer. Use a separate sheet of paper if needed. Marks may be given for a correct method even if the answer is wrong.

Information

- The mark for each question is shown in brackets.
- The maximum mark for this paper is 80.

Name: ....

-5 -4 -3 -2 -1 0 1 2 3 4 5

1. Which average is affected most by an outlier? [1]

....

2. **(a)** Dorcas is thinking of a number. She multiplies it by 4 and then adds 3. She gets the answer −5.

What number was she thinking of? [2]

$x =$

$4x + 3 = -5 \quad -3$
-3
$= -5 - 3.$
$4x = -8 \quad \div 4$

.... −2 .

(b) Ketsia starts with the number 3. She takes away a and then multiplies by b.

She gets an answer of 28. a is a negative integer and b is a positive integer.

Find two possible pairs of values for a and b. [2]

$3 - a \times b = 28 \quad -3$
-3
$-a(b) = 25 \div a$

$a =$ .... $b =$ ....

$a =$ .... $b =$ ....

3. George, Timmy and Ann shared some money in the ratio 5 : 7 : 9

Ann got £32 more than Timmy.

How much money did George get? [3]

$5 : 7 : 9. = \quad 21 = 7.$
$7 \quad 7 \quad 7 \quad \div 3$
Ⓖ $\quad$ Ⓣ $\quad$ Ⓐ $\quad \frac{7}{7}$
$35 : 49 : 63.$

£ .....35.....

4. Jenni plays violin in an orchestra. She has 22 performances this year. For each performance she needs to get a return train ticket which costs £8.85. The orchestra pays $\frac{1}{4}$ of her travel costs.

   (a) Estimate how much Jenni will have to pay in travel costs for the year. **[2]**

   £ ....

   (b) Calculate the exact cost of the train tickets for the year. **[2]**

   £ ....

5. Solve the simultaneous equations $\quad 8x - 5y = 19$ **[4]**
   $$12x + y = 3$$

   $x =$ .... $y =$ ....

6. There are some orange and mint chocolates in a bag. The probability of taking a mint chocolate is $\frac{1}{3}$

Amorreane takes a mint at random and eats it.

The probability of taking a mint chocolate is now $\frac{1}{4}$

How many orange chocolates are in the bag?

[2]

....

7. Increase £340 by 15%

[2]

....

8.

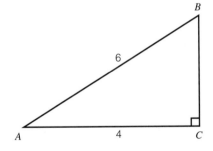

(a) Write down the value of cos A.

[1]

....

(b) What is the value of tan B? Leave your answer as a square root in its simplest form.

[4]

....

9. Work out $\dfrac{5}{8} \div \dfrac{2}{3}$ [1]

....

10. Solve the equation $(x+1)(x-9) = 11$ [4]

....

11. Simplify $\left(3x^2y^5\right)^4$ [2]

....

12. The total TV sales over a five-year period for a national electrical store were 120 000 units. Find the sales figures for computers and TVs in year 4. **[3]**

Year	Computer sales (thousands)	TV sales (thousands)
1	12	
2	14	
3	13	
4	a	$2a + 5$
5	16	

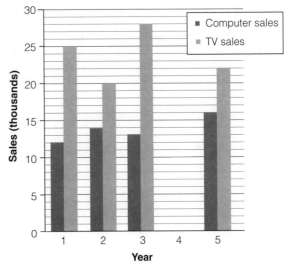

TV sales: ....

Computer sales: ....

13. Find the equation of the line that is parallel to $y = 2x + 7$ and passes through the point $(0, -3)$. **[2]**

....

14. Here is a square and an isosceles triangle.

The length of each of the equal sides of the triangle is 3cm greater than the side of the square.

(a) If the perimeters of the two shapes are equal, find the value of x. **[3]**

....

(b) Show that the height of the triangle is equal to the diagonal of the square. **[3]**

....

....

....

....

15. The graph of $y = \cos x$ is shown for $0° \leqslant x \leqslant 360°$

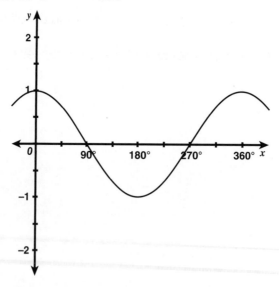

On the same grid, sketch and label the graphs of

(a) $y = -\cos x$ **[2]**

(b) $y = \cos x + 1$ **[2]**

16. Calculate the circumference of this circle.

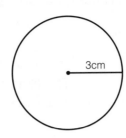

3cm

Leave your answer in terms of π. **[2]**

.... cm

17. Calculate angle *BCD*, giving your reasons. **[3]**

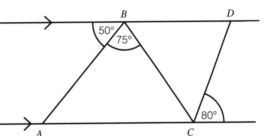

....

....

....

....

18. Work out the next term of this quadratic sequence: **[2]**

–2   3   14   31

....

19. Mike has 240 different films in his collection. He has some films on both Blu-ray and DVD.

The Venn diagram shows the information.

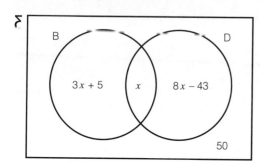

ε = 240 films in the collection
B = films on Blu-ray
D = films on DVD

Mike watches one of his films.

Find the probability that he chooses a film which he has on both Blu-ray and DVD. **[4]**

20. Stereo headphones A cost £a and headphones B cost £b.

When a and b are both increased by £20, the ratio of their prices becomes 5 : 2

When a and b are both decreased by £5, the ratio of their prices becomes 5 : 1

Find the ratio $a : b$ in its lowest terms. **[4]**

21. (a) Write $\sqrt{48}$ as a surd in its simplest form. **[1]**

....

(b) Write $\dfrac{2}{3+\sqrt{7}}$ in the form $a+\sqrt{b}$ **[2]**

....

22. Which is more, 15% of 260 or 18% of 210? **[3]**

Show your working.

....

....

....

....

23. The results of a survey on the lifetime of two different light bulbs are shown on the box plots. **[2]**

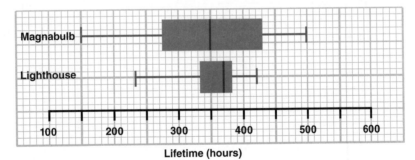

Lifetime (hours)

Give two reasons why Lighthouse bulbs might be better.

Reason 1: ....

....

Reason 2: ....

....

24. Two cones A and B have volumes of 240cm³ and 810cm³.

If the surface area of cone A is 180cm², what is the surface area of cone B?   **[4]**

.... cm²

25. Sketch the graph of $y = x^2 - 8x + 17$, showing clearly the coordinates of the turning point and the coordinates of any intercepts with the coordinate axes.

Write down the equation of the line of symmetry.   **[6]**

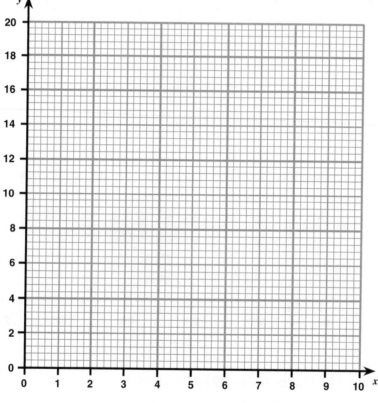

Line of symmetry = ...

GCSE
Mathematics
Higher tier

Paper 2   Time: 1 hour 30 minutes

For this paper you must have:

- a calculator
- mathematical instruments

Instructions

- Use black ink or black ball-point pen. Draw diagrams in pencil.
- Read each question carefully before you start to write your answer.
- Diagrams are not accurately drawn unless otherwise stated.
- Answer **all** the questions.
- Answer the questions in the space provided.
- In all calculations, show clearly how you work out your answer. Use a separate sheet of paper if needed. Marks may be given for a correct method even if the answer is wrong.
- If your calculator does not have a π button, take the value of π to be 3.142 unless the question instructs otherwise.

Information

- The mark for each question is shown in brackets.
- The maximum mark for this paper is 80.

Name: ....

1. A new tablet computer is released and the older model costing £425 is reduced by 12%.

    What is its new price? **[2]**

    £ ....

2. Find two numbers, **greater than 71**, which have the following **two** properties:

    - a multiple of 5 and 7
    - a common factor of 420 and 630 **[4]**

    ....

3. Calculate the area of this shape. Give your answer to 3 significant figures. **[3]**

    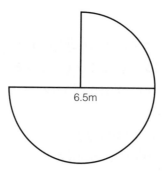

    6.5m

    ....

4. A biologist conducted a survey of different lakes to investigate algae growth and the amounts of nitrate present. She recorded the concentrations of both algae and nitrate in a table.

Algae (cells/ml)	10	12	55	20	5	24	42	44	15	30	22	38
Nitrate (mg/L)	0.5	1.5	3.6	0.7	0.5	1.8	1.5	1.8	1.7	2.3	0.8	2.6

(a) Draw a scatter graph of this data on the grid below. **[2]**

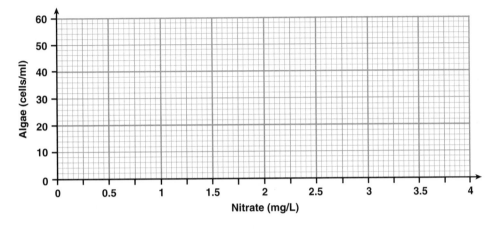

(b) What type of correlation is shown between nitrate levels and algae concentrations? **[1]**

....

(c) Use the graph to estimate the nitrate levels in a lake with an algae concentration of 28 cells/ml. **[2]**

.... mg/L

(d) The biologist wants to use this data to predict algae concentrations for any lake.

Comment on the biologist's idea. **[1]**

....

....

5. Expand

(a) $(2x-1)^2$ [2]

....

(b) $(2x-1)^3$ [2]

....

6. **(a)** Work out the following. Write down all the figures on your calculator display. [2]

$$\frac{\sqrt{35}+6^2}{(2-0.04)^3}$$

....

(b) Round your answer in part (a) to 3 significant figures. [1]

....

7. Show that the point $(-3, -2)$ is on the curve $y = x^2 + 2x - 5$ [2]

....

....

....

....

8. Three feeding bowls, A, B and C, are placed in a cage with two hamsters. The hamsters are equally likely to choose any of the feeding bowls.

Find the probability that both hamsters choose bowl C. **[2]**

....

9. Matthew is estimating the height of his house using the Sun.

A stick 1m long casts a shadow 60cm long. At the same time the shadow of the house is 5.4m long.

How tall is the house? **[3]**

....

10. Sketch the graph of $y = 2^{-x}$ on the axes below. [2]

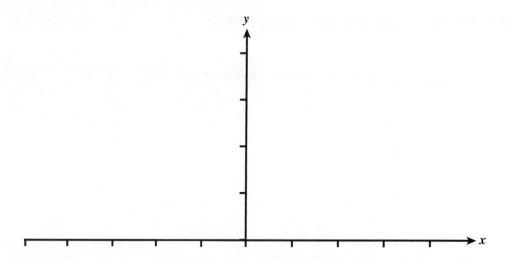

11. The graph shows the temperature of a jug of custard as it cools.

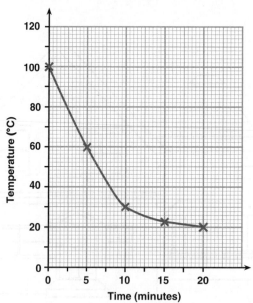

(a) What is the approximate temperature of the room? [1]

....

(b) What is the temperature of the custard after 15 minutes? [1]

....

(c) What is the rate of cooling after 10 minutes? [3]

....

12. Jerome is a salesman. He visits farms to try to sell gates and fencing equipment.

When Jerome visits a farm, the probability that he will make a sale is 0.4

One morning Jerome visits two farms.

(a) Complete the tree diagram to show all the outcomes.   **[2]**

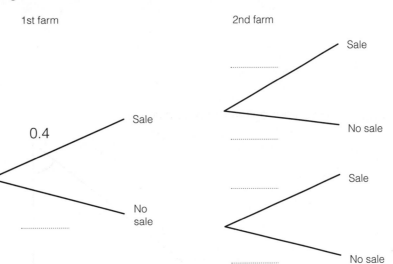

(b) Find the probability that Jerome sells equipment to at least one of the farms.   **[2]**

13. Rearrange $y = \dfrac{10x - 2}{5 - x}$ to make x the subject. [4]

14. Calculate the angle R. Give your answer to 3 significant figures. [3]

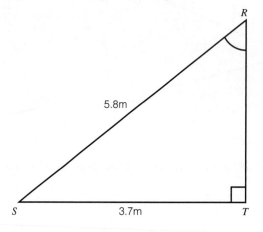

15. Copper and zinc are mixed to form an alloy in the ratio 9 : 7 by mass.

(a) If 27kg of copper are used to make a batch of alloy, how much alloy can be made?   **[3]**

.... kg

(b) The density of copper is 8900kg/m³.

Find its density in g/cm³.   **[4]**

.... g/cm³

(c) If the density of zinc is 7g/cm³, work out the density of the alloy.   **[3]**

.... g/cm³

16. Find the coordinates of the points where the line $y = 2x - 4$ crosses the curve $y = x^2 - 7x + 14$ [4]

17. Prove that $0.0\dot{8}\dot{1} = \dfrac{9}{110}$ [3]

18. Work out the equation of the line that is perpendicular to the line $y = 7 - 2x$ and passes through the point (8, –1).   [4]

19. The ratio of the radii of two cones is 1 : 3

(a) Calculate the ratio of the curved surface areas of the cones. [1]

....

(b) If the volume of the larger cone is 10.8 litres, what is the volume of the smaller cone? [2]

....

20. Find the length *KL*. [3]

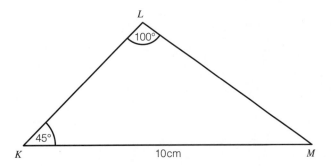

21. The nth term in an arithmetic sequence is given by $U_n = 5n + 2$

Another sequence is given by $V_n = 4U_{n+2} - 2U_n$

Prove that V_n is also an arithmetic sequence.

Is there any value for n such that $U_n = V_n$?

[6]

GCSE
Mathematics
Higher tier

Paper 3   Time: 1 hour 30 minutes

For this paper you must have:

- a calculator
- mathematical instruments

Instructions

- Use black ink or black ball-point pen. Draw diagrams in pencil.
- Read each question carefully before you start to write your answer.
- Diagrams are not accurately drawn unless otherwise stated.
- Answer **all** the questions.
- Answer the questions in the space provided.
- In all calculations, show clearly how you work out your answer. Use a separate sheet of paper if needed. Marks may be given for a correct method even if the answer is wrong.
- If your calculator does not have a π button, take the value of π to be 3.142 unless the question instructs otherwise.

Information

- The mark for each question is shown in brackets.
- The maximum mark for this paper is 80.

Name: ....

1. Lyse and Lysette each test out a biased coin to find out the estimated probability of it landing on heads. Here are the results:

	Number of coin flips	Number of heads
Lyse	40	16
Lysette	80	28

(a) Whose results give the best estimate for the probability of getting heads?

Explain your answer. [1]

....

....

....

....

The coin is flipped 500 times.

(b) How many times do you expect the coin to land on heads? [2]

....

2. Find an integer value of x satisfying $3x + 4 > 19$ and $2x - 1 < 13$ [3]

....

3. Sketch two graphs to show

(a) y directly proportional to x. [2]

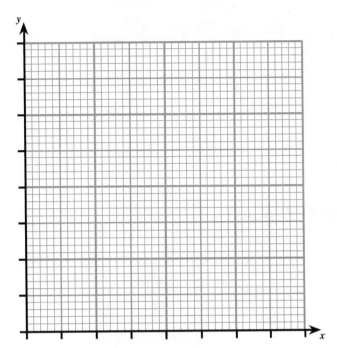

(b) y inversely proportional to x. [2]

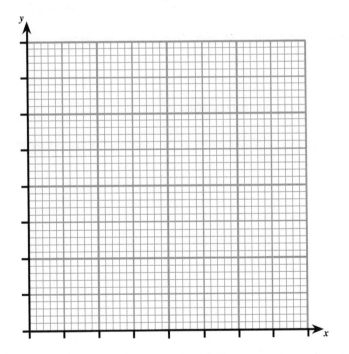

4. Here is a map of part of Norway.

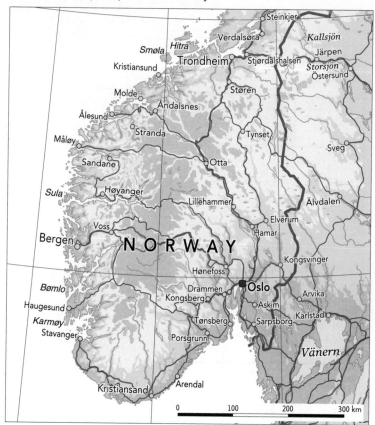

Scale 1 : 6 500 000

(a) Estimate the straight-line distance from Trondheim to Bergen.

[2]

....

(b) The actual road and ferry journey is 700km.

Express the road and ferry distance as a percentage of the straight-line distance.

[2]

....

(c) If it takes 11 hours to make the journey by car, what is the average speed?

[3]

....

5. **(a)** Write $3^5 \div 3^{10}$ as a power of 3. [1]

....

(b) Write 0.000 070 2 in standard form. [1]

....

(c) Evaluate $16^{\frac{1}{2}}$ [1]

....

6. A farmer measures the growth (measured as a change in height) of his crop over a period of eight weeks. The results are shown in the table below.

Growth (cm)	Frequency
$0 \leqslant h < 20$	11
$20 \leqslant h < 40$	38
$40 \leqslant h < 60$	49
$60 \leqslant h < 80$	30
$80 \leqslant h < 120$	12

(a) Which is the modal group? [1]

....

(b) Find the estimated mean growth for the crop over this period. [3]

.... cm

(c) The farmer says that over one-third of his crop grew by more than 60cm during this period.

Does the data support the farmer's claim? [2]

....

....

....

....

....

....

7. The International Space Station (ISS) travels at a speed of 4.48×10^4 km/h.

(a) Write this value as an ordinary number. **[1]**

....

It takes three hours for the ISS to orbit the Earth twice.

(b) Find the distance that the ISS travels in one full orbit around the Earth.

Write your answer in standard form. **[3]**

....

8.

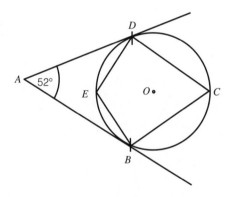

(a) Find angle *BCD*. **[3]**

....

(b) Find angle *BED*. **[2]**

....

9. The value of a motor bike (£V) is given by the formula $V = 18\,000 \times 0.85^t$ where t is the age in whole number of years.

(a) What is the value of the bike when it is new? [1]

....

(b) What is the bike worth after two years? [2]

....

(c) After how many years will its value be below £10 000? [2]

....

10. Solve the equation $\dfrac{6}{x} - \dfrac{4}{2x+2} = 2$, giving your answers to 2 decimal places. [5]

....

11. A hot-air balloon flies from home (H) on a bearing of 080° for 200km to a point D.

At D it turns on to a bearing of 200° for a further 300km to a point E.

Calculate the distance and bearing which will return the balloon home. **[6]**

12. The volume of a cone is given by $\frac{1}{3} \pi r^2 h$

A frustum is the shape left when the top of a cone has been cut off.

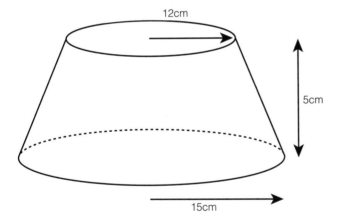

Find the volume of this frustum. Give your answer to 3 significant figures. **[5]**

13. Write $4x^2 - 24x + 41$ in the form $a(x-b)^2 + c$ **[4]**

14. Robson is taking part in a mountain bike race. He must finish in the top 10 to qualify.

The probability that he will qualify if it is dry is 0.8

The probability that he will qualify if it is raining is 0.7

The probability that it will be raining on the day of the race is also 0.7

Find the probability that he will qualify in the race. **[3]**

15. A vicar has one full bottle of communion wine for a midnight mass service.

The bottle holds 750ml of wine to the nearest 5ml.

Each communion cup holds 5ml of wine to the nearest ml.

If he uses the full bottle, what is the smallest number of cups that the vicar will fill for the midnight mass service? **[3]**

16. Functions f and g are given by $f(x) = 3x - 1$ and $g(x) = 1 + 2x$

(a) Find $fg(3)$ **[2]**

(b) Solve the equation $f^{-1}(x)g^{-1}(x) = 4$ **[5]**

17. Evaluate $16^{\frac{3}{4}}$ [2]

....

18. An object travels in a straight line from rest such that its velocity v at time t is shown by the velocity–time graph below.

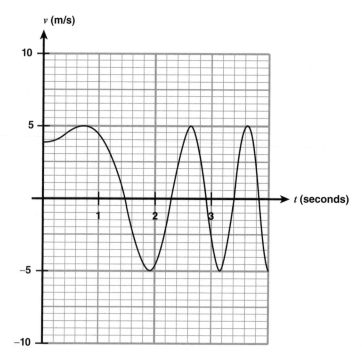

Estimate the acceleration of the object after one second. [5]

.... m/s^2

You are encouraged to show your working out, as you may be awarded marks for method even if your final answer is wrong. Full marks can be awarded where a correct answer is given without working being shown but if a question asks for working out, you must show it to gain full marks. If you use a correct method that is not shown in the mark scheme below, you would still gain full credit for it.

Module 1: Place Value and Ordering

1. (a) –7   (b) 21   (c) –3

   Remember the rules for signs and BIDMAS.

2. (a) False   (b) True   (c) True   (d) False
3. (a) –3°C   (b) –9°C
4. (a) 5.418   (b) 5418   (c) 12 900   (d) 420

   Compare the position of decimal points with the original calculation.

5. (a) £35 [Accept –£35]
   (b) £35 – £14 [1] = £21 [1] [Accept –£21]

Module 2: Factors, Multiples and Primes

1. (a) 1 and 3   (b) 1, 4 and 16   (c) 6, 12, 24 and 48
2. 21 [1] or 45 [1]

   Test the prime numbers between 1 and 50.

3. (a) 55   (b) 7
4. (a) $72 = 2 \times 2 \times 18 = 2 \times 2 \times 2 \times 3 \times 3$ [2] [Two correct steps of decomposition get 1 mark] $= 2^3 \times 3^2$ [1]

   Use factor trees.

   (b) $90 = 2 \times 3 \times 15$ [1] $= 2 \times 3 \times 3 \times 5$ [1] $= 2 \times 3^2 \times 5$ [1]
   (c) $2 \times 3 \times 3$ [1] $= 18$ [1]

   Look for the common factors from the lists in (a) and (b).

   (d) $18 \times 2 \times 2 \times 5$ [1] $= 360$ [1]

   HCF × remaining factors.

5. Bus A: 12   24   36   48   60   72   84
   Bus B: 28   56   84   112   140   168   196 [1]
   LCM = 84 [1]
   8.00am + 84 minutes = 9.24am [1]

Module 3: Operations

1. (a) 53

   Multiplication before addition.

   (b) $\frac{11}{15}$

   Find a common denominator.

   (c) 4

2. $\frac{(-2)^2}{2}$ [1] $= 2$ [1]

3. (a) $\frac{12}{40}$ [1] $= \frac{3}{10}$ [1]   (b) $\frac{50}{18}$ [1] $= 2\frac{7}{9}$ [1]

   Find the reciprocal of the second fraction and multiply.

   (c) $1\frac{15-12}{20}$ [1] $= 1\frac{3}{20}$ [1]   (d) $\frac{3}{2} \times \frac{5}{19}$ [1] $= \frac{15}{38}$ [1]

   Convert to an improper fraction and divide as normal.

4. $1 - \left(\frac{2}{5} + \frac{3}{7}\right)$ [1] $= 1 - \frac{29}{35}$ [1] $= \frac{6}{35}$ [1]

5. $5\frac{1}{4} \times 3\frac{5}{6}$ [1] $= \frac{21}{4} \times \frac{23}{6}$ [1] $= \frac{483}{24} = 20\frac{1}{8}$ m² [1]

6. $1 - \left(\frac{3}{5} + \frac{2}{9}\right)$ [1] $= 1 - \frac{37}{45}$ [1] $= \frac{8}{45}$ [1]

Module 4: Powers and Roots

1. $\sqrt[3]{64}$ $\sqrt[4]{625}$ 5^2 3^3 2^5
2. (a) $x = 7$ (b) $x = 125$ (c) $x = 3$

3. (a) 4^{10}

   Subtract the negative power, i.e. $4^{6-(-4)}$

   (b) $4\sqrt{2}$ (c) $\frac{1}{9}$

   $\sqrt{16} \times \sqrt{2} = 4\sqrt{2}$

4. (a) (i) 3.21×10^5 (ii) 6.05×10^{-4}
   (b) (i) 30×10^{-5} [1] $= 3 \times 10^{-4}$ [1]
   (ii) $41\,000 + 3400 = 44\,400$ [1] $= 4.44 \times 10^4$ [1]

5. (a) $\sqrt{3} \times 4 - \sqrt{3} \times \sqrt{3} = 4\sqrt{3} - 3$
   (b) $8 - 4\sqrt{5} + 12\sqrt{5} - 30$ [1] $= 8\sqrt{5} - 22$ [1]

6. (a) $\frac{4}{\sqrt{7}} \times \frac{\sqrt{7}}{\sqrt{7}} = \frac{4\sqrt{7}}{7}$

   (b) $\frac{5}{2-\sqrt{3}} \times \frac{2+\sqrt{3}}{2+\sqrt{3}}$ [1] $= \frac{5(2+\sqrt{3})}{(2-\sqrt{3})(2+\sqrt{3})} = \frac{10+5\sqrt{3}}{4+2\sqrt{3}-2\sqrt{3}-3}$ [1]
   $= \frac{10+5\sqrt{3}}{1} = 10 + 5\sqrt{3}$ [1]

Module 5: Fractions, Decimals and Percentages

1. $\frac{17}{51}$ $\frac{3}{8}$ 0.475 $\frac{12}{25}$

   Convert all the values to decimals to compare them.

2. $\frac{19}{40}$
3. $5 \div 8 = 0.625$ [1] $7 \div 11 = 0.\dot{6}\dot{3}$ [1]
   $\frac{7}{11}$ is closer to $\frac{2}{3}$ (as $\frac{2}{3} = 0.\dot{6}$) [1]
4. (a) $37 \times 49 = 1813$ [1]
   So $3.7 \times 4.9 = 18.13$ [1]
   (b) $3696 \div 14$ [1] $= 264$
   So $369.6 \div 1.4 = 264$ [1]
5. 34×129 [1] $= 4386$ [1]
   So total cost is £43.86 [1]

   Use long multiplication.

6. $\frac{4}{9}$
7. $99x = 45$ [1] $x = \frac{45}{99} = \frac{5}{11}$ [1]
8. $990x = 531$ [1] $x = \frac{531}{990} = \frac{59}{110}$ [1]

Module 6: Approximations

1. (a) 9.49   (b) 6.554   (c) 5.60

   Use a calculator first.

2. (a) 400 000   (b) 0.0497   (c) 3.142

   Use calculator to find π.

3. $500 \div 35$ [1] $= 14.3g$ [1]
4. $70 \times 40 \times 60$ [1] $= 168\,000$ [1]
   £1680 [1]
5. Upper bound = 1.735m [1]
   Lower bound = 1.725m [1]

   Half a centimetre above and below 1.73

6. 500.5×8 [1] $= 4004$ml [1]

   Use the upper bound to find the maximum amount.

7. Minimum value $= \frac{6.75}{0.345}$ [1] $= 19.57$ [1]
   Maximum value $= \frac{6.85}{0.335}$ [1] $= 20.45$ [1]

Module 7: Answers Using Technology

1. **(a)** 410.0625 **(b)** 7 **(c)** 5

2. $\dfrac{19.21}{32.76}$ **[1]** = 0.586 385 836 **[1]**

   Do not round at any point.

3. 1.2×10^{-9}

4. **(a)** $2\dfrac{20}{21}$

   Your calculator might show

   (b) $31\dfrac{1}{18}$

5. **(a)** $1.496 \times 10^8 \div 400$ **[1]** $= 374\,000$ **[1]** $= 3.74 \times 10^5\,\text{km}$ **[1]**

   (b) $1.392 \times 10^6 \div 3.48 \times 10^3$ **[1]** $= 400$ **[1]** $= 4 \times 10^2$ **[1]**

   (c) The Sun is 400 times bigger and 400 times further away (this is why a total eclipse appears the way it does).

   (d) $1.496 \times 10^8 + 3.74 \times 10^5$ **[1]** $= 149\,974\,000$ **[1]**
   $= 1.5 \times 10^8\,\text{km}$ **[1]**

Module 8: Algebraic Notation

1. **(a)** $5a + 2$ **(b)** $15h + 2k$ **(c)** $-2a + 2b$ **(d)** $8x^2 - x + 1$

2. **(a)** q **(b)** $27p^{12}$ **(c)** $6f^5g$ **(d)** $\dfrac{1}{8p^6}$

3. **(a)** $12k^2$ **(b)** a^3 **(c)** $p^{\frac{5}{2}}$

   (d) $p^{\frac{13}{2}}$ or $\sqrt{p^{13}}$

4. **(a)** $\dfrac{1}{64}$

   Remember to deal with the negative index first by using 'one over...'. So $4^{-3} = \dfrac{1}{4^3}$

   (b) $\dfrac{25}{4}$ **(c)** 4 **(d)** 1

5. **(a)** Equation **(b)** Identity **(c)** Identity

   (d) Equation **(e)** Equation

Module 9: Algebraic Expressions

1. $8x^2 - 20xy + 6xy - 15y^2$ **[1]** $= 8x^2 - 14xy - 15y^2$ **[1]**

2. $p(4p^2 - 1)$ **[1]** $= 4p^3 - p$ **[1]**

   You can multiply the first two terms or the final two terms first; you should still end up with the right answer.

3. $6x - 3 + 4x + 32 + 5$ **[1]** $= 10x + 34$ **[1]** $= 2(5x + 17)$ **[1]**

4. $(a - 16)(a - 3)$ **[2]**

5. $\dfrac{(x-9)(x+2)}{(x+9)(x-9)}$ **[2]** $= \dfrac{x+2}{x+9}$ **[1]**

   In questions like these, expect a factor to cancel. So if the denominator is $(x + 9)(x - 9)$, you can expect a factor in the numerator to be either $x + 9$ or $x - 9$.

6. $\dfrac{(a+3) - (a+2)}{(a+2)(a+3)}$ **[1]**

   Don't forget to put brackets round the second term in the numerator here. An easy thing to forget.

   $= \dfrac{1}{(a+2)(a+3)}$ **[1]**

7. $\dfrac{-x-4}{x+1}$ **[1 for top of fraction and 1 for bottom of fraction]**

8. $(x+2)(x-3)(x+4) = (x+2)(x^2 + x - 12)$

   $= x(x^2 + x - 12) + 2(x^2 + x - 12)$

   $= x^3 + x^2 - 12x + 2x^2 + 2x - 24$

   $= x^3 + 3x^2 - 10x - 24$ **[1 for each correct term]**

9. $\dfrac{x-9}{6} \div \dfrac{x^2 - 9x}{3} = \dfrac{x-9}{6} \times \dfrac{3}{x^2 - 9x}$ **[1]** $= \dfrac{x-9}{6} \times \dfrac{3}{x(x-9)}$ **[1]**

   $= \dfrac{1}{2x}$ **[1 for $\dfrac{1}{2}$ and 1 for x in denominator]**

10. $(3x - 5)(x + 8)$ **[2]**

11. $(2x + 1)^3 = (2x + 1)(4x^2 + 4x + 1)$ **[1]**

    $= 8x^3 + 12x^2 + 6x + 1$ **[2] [−1 for an error or omission]**

12. $\dfrac{x^2 - 5x - 84}{x^2 + 5x - 14} = \dfrac{(x+7)(x-12)}{(x+7)(x-2)}$ **[2]** $= \dfrac{x-12}{x-2}$ **[1]**

Module 10: Algebraic Formulae

1. **(a)** 36 **(b)** 48 **(c)** $2\dfrac{2}{3}$

   (d) $-7\dfrac{2}{3}$ **(e)** 6

2. $xy - 8y = 2 + 3x$ **[1]**

   $xy - 3x = 2 + 8y$

   $x(y - 3) = 2 + 8y$ **[1]**

   $x = \dfrac{2 + 8y}{y - 3}$ **[1]**

3. $s - ut = \dfrac{1}{2}at^2$ **[1]**

   $a = \dfrac{2(s - ut)}{t^2}$ **[1]**

4. $y - 5 = \dfrac{2}{x}$ **[1]**

   $x = \dfrac{2}{y - 5}$ **[1]**

5. $q = r - p$ **[2]**

   Try to write the given information mathematically first. This question is basically telling you that $p + q = r$, then asking you to rearrange the equation.

6. **(a)** 31.1 degrees Celsius (1 d.p.)

   (b) $9T_C = 5T_F - 160$ **[1]**

   $T_F = \dfrac{9T_C}{5} + 32$ **[1]**

   (c) 140 degrees Fahrenheit

Module 11: Algebraic Equations

1. $5x - 10 - 3x + 12 = 4$ **[1]**

   $2x = 2$

   $x = 1$ **[1]**

2. $x^2 - 3x - 28 = 0$

   $(x + 4)(x - 7) = 0$ **[1]**

   $x = -4$ **[1]**, $x = 7$ **[1]**

3. $(3x - 5)(x + 2) = 0$ **[1]**

   $x = \dfrac{5}{3}$ **[1]**, $x = -2$ **[1]**

4. $x = 5$ **[1]**, $y = 2$ **[1]**

5. $(x - 4)^2 - 20 = 0$ **[2]**

   $x = 4 \pm \sqrt{20}$ **[1]**

6. $3\left[x^2 + 8x + \dfrac{40}{3}\right]$ **[1]** $= 3\left[(x+4)^2 - \dfrac{8}{3}\right]$ **[1]** $= 3(x+4)^2 - 8$ **[3]**

7. $x = \dfrac{8 \pm \sqrt{220}}{6}$ **[1]**

   $x = -1.14$ **[1]**, $x = 3.81$ **[1]**

8. $x^2 - x - 2 = 2x + 2$ **[1]**

   $x^2 - 3x - 4 = 0$

   $(x + 1)(x - 4) = 0$ **[1]**

   $x = -1$, $y = 0$ **[1]**

   $x = 4$, $y = 10$ **[1]**

   Distance $= \sqrt{5^2 + 10^2} = \sqrt{125}$ **[1]**

9. $8 - x^2 = x - 4$

$x^2 + x - 12 = 0$ **[1]**

$(x + 4)(x - 3) = 0$ **[1]**

$x = -4$ **[1]**, $x = 3$ **[1]**

$y = -8$ **[1]**, $y = -1$ **[1]**

10. $2x^2 - 11x + \dfrac{9}{2} = 0$ **[2]**

$2x^2 - 10x + 7 = x + \dfrac{5}{2}$ **[2]**

So need to plot $y = x + \dfrac{5}{2}$

Remember to start with the equation you are trying to **solve**, and add or subtract terms from both sides to obtain the equation you **have**.

11. List of at least five iterations **[2]** leading to $x = 2.49$ **[1]**

Module 12: Algebraic Inequalities

1. $7x > 35$ **[1]** $\qquad x > 5$ **[1]**

2. **(a)** $x \leqslant -1$ **(b)** $x > 3$ **(c)** $-4 \leqslant x < 2$

3. $-3, -2, -1, 0, 1, 2, 3$ **[2]**

4. $(x - 8)(x - 1) \leqslant 0$ **[1]**

[1, 8] **[1 for brackets and 1 for correct values]**

5. **(a)**

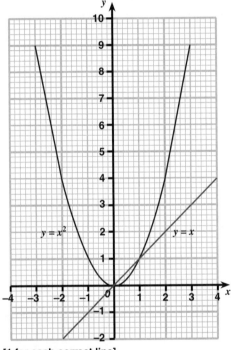

[1 for each correct line]

(b) $0 < x < 1$

(c) $(-\infty, 0) \cup (1, \infty)$ **[2]**

Even if the sketch was not asked for in this question, it would be useful to draw one.

6. $2 \leqslant x \leqslant 5$ and $1 \leqslant y \leqslant 6$ **[2 for correct values and 2 for correct boundaries]**

7.

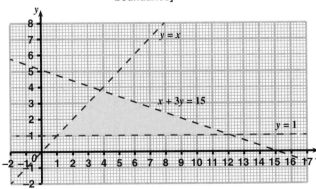

[1 for each correct (dashed) line and 1 for triangle shaded]

In questions like this, rearrange each equation into the form $y = mx + c$. It will make each line easier to sketch or plot.

Module 13: Sequences

1. 5, –4 **[1 for each correct term]**

2. 44, 57 **[1 for each correct term]**

3. 1, 3, 13, 63, 313 **[2 if fully correct; 1 if one error]**

4. **(a)** $U_n = 4n + 7$ **[1 for 4n; 1 for 7]**

(b) Yes **[1]** since there is a whole number solution to $4n + 7 = 303$
$(n = 74)$ **[1]**

5. **(a)** $a + b + c = 7$

$4a + 2b + c = 6$ **[1]**

$9a + 3b + c = 7$

$\left. \begin{array}{l} 5a + b = 1 \\ 3a + b = -1 \end{array} \right\}$ **[1]**

$a = 1$, $b = -4$, $c = 10$ **[1 for each correct value]**

Always write down the three equations, then subtract the second from the third, and the first from the second.

(b) $U_{20} = 330$

6. **(a)** $U_n = \left(2\sqrt{3}\right)^n$ **[2]** **(b)** $U_3 = 24\sqrt{3}$

Module 14: Coordinates and Linear Functions

1. Line A: 6 **[1]** Line B: –3 **[1]** Line C: $\dfrac{-2}{3}$ **[1]** Line D: 1 **[1]**

2. Lines A and C are parallel. **[1]**
Lines B and D are perpendicular. **[1]**

3. $x = 0$

4. $y = 9 - 6x$ **[1 for 9; 1 for –6x]**

5. $y = 5x + c$ **[1]**

$1 = -15 + c$ **[1]**

$c = 16$

$y = 5x + 16$ **[1]**

6. $m = \dfrac{-5 - (-10)}{6 - (-4)}$ **[1]**

$m = \dfrac{1}{2}$

$y = \dfrac{1}{2}x + c$ **[1]**

Substitute in one coordinate to give $c = -8$ **[1]**

$y = \dfrac{1}{2}x - 8$ **[1]**

Always show your working out for the gradient. Even if the final answer is wrong, you will still pick up some marks.

7. $y = \dfrac{1}{3}x + 5$ **[1]**

So need $m = -3$

$y = -3x + c$ **[1]**

$-4 = -15 + c$ **[1]**

$c = 11$ **[1]**

$y = 11 - 3x$ **[1]**

Module 15: Quadratic Functions

1. $x = \dfrac{-b}{2a} = \dfrac{-(-30)}{10}$ **[1]**

$x = 3$ **[1]**

2. $(x + 4)(x - 11) = 0$ **[1]**

$x + 4 = 0$ or $x - 11 = 0$

$x = -4$ **[1]** or $x = 11$ **[1]**

3. $(3x - 2)(x + 4) = 0$

$3x - 2 = 0$ or $x + 4 = 0$ **[1]**

$x = \dfrac{2}{3}$ **[1]** or $x = -4$ **[1]**

4. **(a)** $x^2 + 6x + 4 = (x+3)^2 - 5$ **[2]**

So minimum point occurs at $(-3, -5)$ **[1]**

(b) $(0, 4)$

5. **(a)** $(x-2)(x-4) - 48 = x^2 - 6x - 40$ **[1]**

$= (x+4)(x-10)$ **[1]**

So crosses axes at $x = -4$ and $x = 10$ **[2]**

(b) $(0, -40)$

(c) $x = 3$

6. **(a)** $3x^2 - 12x + 22 = 3\left(x^2 - 4x + \dfrac{22}{3}\right)$ **[1]**

$= 3\left((x-2)^2 - 4 + \dfrac{22}{3}\right)$ **[1]**

$= 3\left((x-2)^2 + \dfrac{10}{3}\right) = 3(x-2)^2 + 10$ **[3]**

(b) $(2, 10)$

> This is a common type of question. Always factor out the x^2 coefficient completely; complete the square, then multiply through by the number again.

7. $25 + 10x - x^2 = -\left(x^2 - 10x - 25\right)$ **[1]**

$= -\left((x-5)^2 - 50\right)$ **[1]** $= 50 - (x-5)^2$ **[1]**

Maximum point occurs at $(5, 50)$ **[1]**

> Even if the number before the x^2 term is negative, still factor it out completely as in the first line here.

Module 16: Other Functions

1. **(a)** $y = 5x + 3$

$x = \dfrac{y-3}{5}$ **[1]**

$f^{-1}(x) = \dfrac{x-3}{5}$ **[1]**

(b) $ff(x) = 5(5x+3) + 3$ **[1]** $= 25x + 18$ **[1]**

2.

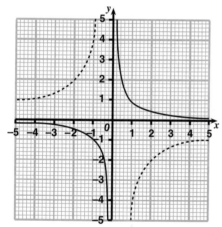

Solid curve $y = \dfrac{1}{x}$ **[1 for correct shape and 1 for correct position]**

Dashed curve $y = \dfrac{-4}{x}$ **[1 for correct shape and 1 for correct position]**

> When sketching reciprocal graphs, make sure the curves never touch or cross the axes.

3. **(a)** –4   **(b)** –3   **(c)** $\dfrac{-1}{2} - (-2)$ **[1]** $= \dfrac{3}{2}$ **[1]**

4.

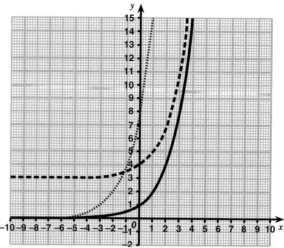

(a) See solid line above. **[1 for graph shape and 1 for line passing through the point (0, 1)]**

(b) See dashed line above. **[1 for graph shape and 1 for line passing through the point (0, 4)]**

(c) See dotted line above. **[1 for graph shape and 1 for line passing through the point (0, 8)]**

> If the curve you are sketching crosses any axes, always make clear where it does so.

5. A translation of 90° in the direction of the positive x-axis (to the right) **[2]** and a translation of 2 units in the direction of the positive y-axis (up) **[2]**.

6. $y = \dfrac{3x+1}{1-x}$

$y - xy = 3x + 1$ **[1]**

$y - 1 = 3x + xy$ **[1]**

$y - 1 = x(3 + y)$ **[1]**

$x = \dfrac{y-1}{3+y}$ **[1]**

$P^{-1}(x) = \dfrac{x-1}{3+x}$ **[1]**

7. $x^2 - 2x + 5 = (x-1)^2 + 4$ **[2]**

Translation of $y = x^2$ **[1]** 1 unit to the right **[1]** followed by translation **[1]** 4 units up **[1]**.

> This could also be described as a single translation through the column vector $\begin{pmatrix} 1 \\ 4 \end{pmatrix}$

Module 17: Problems and Graphs

1. $x^2 + y^2 = \left(\sqrt{2}\right)^2$

so shape is a circle **[1]**, centre (0, 0) **[1]** of radius $\sqrt{2}$ **[1]**

2. **(a)** $x^2 + y^2 = \dfrac{9}{4}$

$x^2 + y^2 = \left(\dfrac{3}{2}\right)^2$ **[1]**

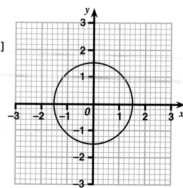

[2]

(b) Equation of tangent is $\frac{1}{2}x + \sqrt{2}y = \frac{9}{4}$ **[2]**

$2x + 4\sqrt{2}y = 9$ **[1]**

$2x + 4\sqrt{2}y - 9 = 0$ **[1]**

3. (a) Area of triangle 1 is $\frac{1}{2} \times \frac{1}{2} \times 5 = \frac{5}{4}$

Area of trapezium 2 is $\frac{1}{2} \times \frac{(5+8)}{2} = \frac{13}{4}$

Area of trapezium 3 is $\frac{1}{2} \times \frac{(8+9)}{2} = \frac{17}{4}$

Area of trapezium 4 is $\frac{1}{2} \times \frac{(8+9)}{2} = \frac{17}{4}$

Area of trapezium 5 is $\frac{1}{2} \times \frac{(5+8)}{2} = \frac{13}{4}$

Area of triangle 6 is $\frac{1}{2} \times \frac{1}{2} \times 5 = \frac{5}{4}$

[4 for calculating areas of the six strips]

The total area is therefore

$\frac{5}{4} + \frac{13}{4} + \frac{17}{4} + \frac{17}{4} + \frac{13}{4} + \frac{5}{4} = \frac{70}{4} = \frac{35}{2}$ units^2 **[2]**

Show clearly how you have calculated each separate area. You are less likely to make a mistake this way.

(b) Draw a suitable tangent line at (0, 0): **[2]**

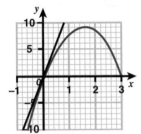

Choose any point lying on the tangent line, e.g. $\left(\frac{1}{2}, 6\right)$ **[1]**

Gradient from (0, 0) to $\left(\frac{1}{2}, 6\right)$ is 12 **[1]**

Equation of tangent is therefore $y = 12x$ **[1]**

4. Substituting (h, k) into the equation $x^2 + y^2 = a^2$ gives $h^2 + k^2 = a^2$ **[1]**

The radius line from (0, 0) to (h, k) has gradient $\frac{k}{h}$ **[1]**

The gradient of the tangent is therefore $\frac{-h}{k}$ **[1]**

The equation of the tangent is therefore $y = \frac{-h}{k}x + c$ **[1]**

Substituting $x = h$ and $y = k$ gives $k = \frac{-h^2}{k} + c$ **[1]**

So $c = \frac{h^2}{k} + k = \frac{h^2 + k^2}{k}$

So $c = \frac{a^2}{k}$

Therefore equation of tangent is $y = \frac{-h}{k}x + \frac{a^2}{k}$ **[1]**

Or $hx + ky = a^2$

Module 18: Converting Measures

1. 0.85×1.25 (or 850×1250) **[1]**
= 1.06m^2 (3 s.f.) (or 1062500mm^2) **[1]**

2. 3.4×1000^2 **[1]** = 3 400 000m^2 **[1]**

3. $100 \times 100 \times 100$cm^3 = 1m^3 **[1]**
1 000 000g = 1000kg **[1]**

4. $68 \div 5 \times 8$ or $110 \div 8 \times 5$ **[1]**
108.8 or 68.75
110km/h is faster **[1]**

5. $24 \times 60 \times 60$m/hour **[1]** $\div 1000$ **[1]** = 86.4km/h **[1]**
Multiply m/s by 60×60 to convert to m/hr.

6. (a) Bob 626 $\div$ 35 = £17.89/h **[1]**
Leila 30 000 $\div$ (32 $\times$ 52) **[1]** = £18.03/h
Shahida has best hourly rate. **[1]**

(b) Shahida $18.4 \times 30 \times 52$ = £28 704
Bob 626×52 = £32 552
[1 for either Bob or Shahida correct]
Bob earns most per year. **[1]**

1 year = 52 weeks

7. (a) $463 \div 24$ **[1]** = 19.29g/cm^3 (2 d.p.) **[1]**
(b) $19.29\ldots \times 30$ **[1]** = 578.75g **[1]**
(c) $(463 + 578.75) \times 23.50$ **[1]** = £24 481.13 **[1]**
[No mark for 24 481.125]

Module 19: Scales, Diagrams and Maps

1. (a) $3 \times 20 = 60$m
(b) $57 \div 20$ **[1]** = 2.85m **[1]**

Scale 1 : n means 1 and n given in cm. Convert n to m or km as necessary.

2. (a) 5.3cm to 5.5cm
(b) $5.3 \times 5 000 000$ **[1]** = 265km **[1]** **[Accept up to 275km for a 5.5cm measurement]**
(c) $203°$ ($\pm3°$) **[2]**
[1 mark for correct construction or 157°]
(d) $(203) - 180$ OR $(203) + 180 - 360$ **[1]** = $023°$ **[1]**

3. (a)

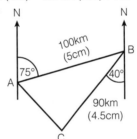

[1 for correct drawing; 1 for accuracy]
(b) $318°$ ($\pm3°$) **[1]**
58km (±2km) **[1]**

All bearings are measured clockwise from North line.

4. (a) $18 \times 25 000$ **[1]** = 450 000cm = 4.5km **[1]**
(b) **[2]**

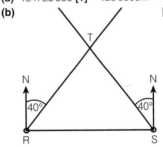

[1 for one bearing correct]

Module 20: Comparing Quantities

1. $\frac{14}{5}$ 37.5% $3\frac{5}{8}$

[2 if fully correct; 1 for two correct]

2. 20%   0.21   0.211 $\frac{2}{9}$ $\frac{3}{10}$

[2 if fully correct; 1 for four in correct order]

Change all amounts to decimals to find the correct order.

3. $2.8 - 2.5 = 0.3$
$2.5 - 2.125 = 0.375$ **[1]**
$0.3 < 0.375$ so $2\frac{4}{5}$ closer **[1]** **[or equivalent working in fractions]**

4. $\frac{3}{8} \times \frac{5}{6}$ [1] $= \frac{5}{16} = 31.25\%$ [1]

Or $\frac{5}{6}$ of 12 = 10 [1]

$\frac{10}{32} = 31.25\%$ [1]

5. (a) 2.54 ÷ 3.45 OR 5.99 ÷ 3.45 [1] → 73.6% (1 d.p.) [1]
   (b) 3.45 × 1.15 [1] (= £3.97)
       (3.45 × 1.15) × 80 [1] (= £317.40)
       55 × 6.99 + 25 × 4 [1] = £484.45
       £484.45 − £317.40 = £167.05
       167.05 ÷ 317.40 = 52.6% [1]

   Percentage profit = (profit ÷ original amount) × 100

6. (a) 0.4 × 0.55 [1] + 0.65 × (1 − 0.55) [1] = 51.25% [1]
   (b) $\frac{35}{100}$ [1] $= \frac{7}{20}$ [1]

Module 21: Ratio

1. 20cm : 3m [1] simplifies to 2 : 30 [1] All the rest simplify to 2 : 3

   Divide all parts of the ratio by a common factor.

2. (a) 1 + 2 + 7 [1] = 10
       2 litres ÷ 10 [1] = 200ml
       Navy blue: 200ml; grey: 400ml; white: 1400ml [1]
   (b) 1200 ÷ 2 = 600ml and 6000 ÷ 7 = 857ml so grey is limiting colour [1]
       1 lot is 600ml so total 600 × 10 = 6000ml = 6 litres [1]

3. A is $\frac{5}{8}$ of B

4. (a) 50 ÷ 5 = 10 so 3 biscuits need 10g.
       3 × 7 = 21 biscuits
       [1 for any correct method; 1 for correct answer]
   (b) 175 ÷ 5 = 35 so 35g flour for 3 biscuits.
       35 × 4 = 140g
       [1 for any correct method; 1 for correct answer]

5. No
   Blue costs 30 ÷ 5 [1] = £6/litre and yellow 28 ÷ 7 = £4/litre
   Cost for 7 litres 5 × 4 + 2 × 6 [1] = £32
   Lucy gets £4.50 × 7 = £31.50. So she loses 50p [1]

Module 22: Proportion

1. 59 × 2.5 [1] = 147.5kcal [1]
2. (a) 500 × 1.29 = €645
   (b) (645 − 570) ÷ 1.33 [1] = £56.39 [1]

   Each £ receives €1.29 → multiply. Each €1.33 receives £1 → divide.

3. (a) 1360 ÷ 80 [1] = 17 litres/cow
       17 × (80 + 25) [1] = 1785
       1785 × 365 × 0.3 [1] = £195 457.50 [1]
   (b) 800 ÷ 105 = 7.6 days
       [1 for any correct method; 1 for correct answer]

   6 tonnes is 800 cattle feeds.

4. Scale factor × 1.5 [1]
   PR = 6 ÷ 1.5 [1] = 4cm
   TU = 4.5 × 1.5 = 6.75cm
   [1 for both correct values]

   Scale factor = $\frac{9}{6}$ = 1.5

5. $\frac{270}{120} = \frac{9}{4}$ so scale factor for length = $\frac{3}{2}$ [1] and for

   volume = $\frac{27}{8}$

   Volume of P = 2700 ÷ $\frac{27}{8}$ [1] = 800cm³ [1]

Module 23: Rates of Change

1. (a) 10 000 × 1.02⁴ or other stepwise method [1]
       = £10 824.32 [1]
       [1 for 10 000 × 1.02]

   (b) 10 000 × 1.15 [1] × 1.06 × 0.82 × 1.01 [1]
       = £10 095.76 [1]

2. No [1]
   6500 × 1.03 = 6695 so £195 is correct for first year but the money invested for year 2 is 6500 + 195 so it will earn more interest [1].
   Actual interest = 6500 × 1.03³ − 6500 = £602.73 [1]

3. (a) 20cm   (b) Between 4 and 8 minutes   (c) 20 ÷ 4 = 5cm/min

   Steepest gradient gives fastest speed.

4. (a) 210 ÷ 10 [1] = 21m/s (±0.4m/s) [1]
   (b) Drawing tangent at t = 5 [1]
       Attempt at gradient [1]
       13m/s (±2m/s) [1]

Module 24: Constructions

1. [1 for correct construction lines shown; 1 for accuracy]

   Always leave your construction lines and arcs, the examiner needs to see them.

2. [1 for width 3.5cm; 1 for accurate lengths and 90° angles]
3. [1 for correct construction; 1 (conditional on construction) for angle of 90° (±2°)]

4.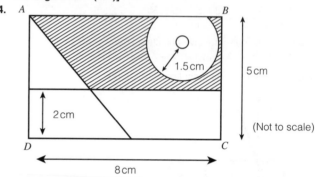

   [1 for circle radius 1.5cm round tree; 1 for line 2cm from DC; 1 for correct area shaded]
5. [1 for equilateral triangle used to construct 60°; 1 for construction of angle bisector; 1 (conditional on construction) for angle of 30° (±2°)]

Module 25: Angles

1. For tessellation, the angles must sum to 360°.
   Internal angle of a regular pentagon = 108°
   360 ÷ 108 is not a whole number.
   [1 for 108°; 1 for written or diagrammatic explanation]

2. 180 − 157.5 [1] = 22.5
   360 ÷ 22.5 = 16 [1]

   It is always the **external** angles that sum to 360°.

3. Internal angle of hexagon = 120°
   Internal angle of octagon = 135°
   [1 for one correct method]
   360 − (120 + 135) [1] = 105° [1]

4. Triangle ABC isosceles
   Angle ACB = 180 − 2 × 70 [1] = 40°
   Angle FCD = 40° (opposite angles)
   Angle FED = 360 − (130 + 130 + 40) = 60° [1]
      (angles in a quadrilateral sum to 360°)
   [1 for all four underlined reasons]

   Give a reason for each step of your working.

5. e.g. ∠BPQ = 65° (opposite angles equal)
   ∠PRQ = 180 − 120 = 60° (angles on straight line sum to 180°)
   ∠CPR = 60° [1] (alternate angles are equal)
   [1 for any correct, complete reasons]

Module 26: Properties of 2D Shapes

1. A – equilateral, B – obtuse isosceles, C – obtuse scalene, D – right-angled isosceles, E – scalene
   [3 if fully correct; 2 if three correct; 1 if one correct]

2. $\angle QBC = c$ (alternate angles are equal) **[1]**
   $\angle PBA = a$ (alternate angles are equal) **[1]**
   $a + b + c = 180°$ (angle sum on straight line = 180°) **[1]**
   Therefore angles in a triangle sum to 180°.

   > A proof must work for every possible value so you will need to use algebra.

3. All angles at centre are equal and 360 ÷ 8 = 45°
   All triangles isosceles so base angle = (180 − 45) ÷ 2 **[1]** = 67.5°
   1 internal angle = 2 × 67.5 **[1]** = 135°
   Total = 135 × 8 = 1080° **[1]**

4. Kite

5. Internal angle of an octagon = 135°
   Internal angle of an equilateral triangle = 60° **[1 for either 135 or 60]**
   Angles at a point = 360° so x = 360 − (135 + 90 + 60) **[1]** = 75° **[1]**

Module 27: Congruence and Similarity

1. B and E
2. G and I

   > Congruent = same shape **and** size

3. $\angle PQS = \angle QSR$ (alternate angles) **[1]**
   $\angle PSQ = \angle SQR$ (alternate angles) **[1]**
   QS common to both triangles, so congruent (ASA) **[1]**
4. Yes, both right angled with sides 3, 4, 5cm **[1]**, so SSS (or RHS) **[1]**
5. Scale factor for length = × 1.5 **[1]** so scale factor for volume = 1.5^3 **[1]**
   $320 × 1.5^3$ **[1]** = 1080cm³ **[1]**

   > (Scale factor for length)² = scale factor for area
   > (Scale factor for length)³ = scale factor for volume

Module 28: Transformations

1. **(a)** Reflection **[1]** in $y = x + 1$ **[1]**
   (b) Reflection in $x = −1$ **[2 if fully correct; 1 if line drawn only]**
2. Enlargement **[1]**, scale factor $\frac{1}{2}$ **[1]**, centre (0, 0) **[1]**

   > Write one fact for each mark.

3.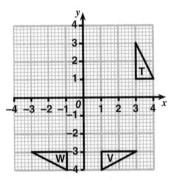
   (a) Correct plot
   (b) **[2 if fully correct; 1 for rotation]**

   > Make sure you do the transformation on the correct shape.

   (c) **[2 if fully correct; 1 for reflection of V if it was incorrect in part (b)]**
   (d) Reflection **[1]** in line $y = −x$ **[1]**
4.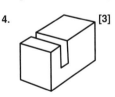
   (a) **[2 if fully correct; 1 for correct scale factor × 2]**

(b) **[2 if fully correct; 1 for correct scale factor × 0.5]**
(c) **[2 if fully correct; 1 for correct scale factor × −1]**

Module 29: Circles

1. **(a)** radius   **(b)** diameter
   (c) tangent   **(d)** circumference
   (e) arc   **(f)** sector
   (g) segment   **(h)** isosceles
2. **(a)** False. If PR was a diameter then angle at Q would be 90°.

   > If you do not give a correct reason you will not get a mark, even if you have the correct answer.

   (b) True. Opposite angles in a cyclic quadrilateral sum to 180°.
   (c) False. $\angle SPR = 180 − (50 + 86)$ **[1]** = 44°
   If SV is a tangent then angle RSV = angle SPR (alternate segment theorem) **[1]**.
3. **(a)** $\angle PRS = 58°$ **[1]**. Angles at the circumference on the same arc are equal **[1]**.
   (b) $\angle QPR = 24°$ (angles at the circumference on the same arc are equal) **[1]**
   $\angle PQR = 90°$ (angle on a diameter)
   $\angle PRQ = 180 − (90 + 24) = 66°$ (angles in a triangle sum to 180°) **[1]**
   $\angle PRQ = 66°$ **[1]**
4. **(a)** **(i)** $AX = DX$ so triangle DXA isosceles.
   $\angle DAX = (180 − 46) ÷ 2$ **[1]** = 67°
   $\angle ABD = \angle DAX = 67°$ (alternate segment theorem) **[1]**
   (ii) $\angle ADX = 67°$ and $\angle BDY = 67°$ (alternate angles) **[1]**
   $\angle BDA = 180 − (2 × 67) = 46°$ (angles on a straight line sum to 180°) **[1]**
   (b) $\angle BAD = 180 − (67 + 46) = 67°$ **[1]**
   Therefore isosceles, as two angles equal **[1]**.
   (c) $\angle CFD = \angle BFA$ (opposite angles) **[1]**
   $\angle ACD = \angle ABF = 67°$ (angles on same arc are equal)
   Angles in both triangles are the same, so they are similar **[1]**.

Module 30: Properties of 3D Shapes

1. Pentagonal-based **[1]** pyramid **[1]**
2. A square-based pyramid **[1]** and triangular prism **[1]** accurately represented by diagrams **[1]**
3.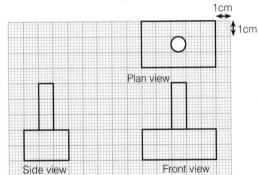

   Plan view

   Side view   Front view

   [1 for each drawing; 1 for accuracy]

4. **[3]**

Module 31: Perimeter, Area and Volume

1. **(a)** Area: 8 × 3 **[1]** = 24m² **[1]**
   Perimeter: 26m **[1]**
   (b) Area: $\frac{1}{2}$ × 14 × 3 **[1]** = 21cm² **[1]**
   Perimeter: 22.6cm **[1]**
2. 200 × 80 − 25 × 20 **[1]** = 15 500cm² **[1]** (or equivalent calculation to 1.55m²)
3. π × 9 **[1]** = 28.274… **[1]** = 28cm **[1]**

   > Make sure you know (and use) the correct formula.

4. $\pi \times 4^2$ **[1]** $= 16\pi$ **[1]** cm² **[1]**

   Here the diameter is given but you need the radius to work out area.

5. $\dfrac{\pi \times 5}{2} + 8 + \dfrac{\pi \times 5}{2} + 8$ (or equivalent) **[1]** $= 5\pi + 16 = 31.7$cm **[1]**

6. $(\pi \times 4^2 \times 2)$ **[1]** $+ (\pi \times 8 \times 5)$ **[1]** $= 32\pi + 40\pi = 72\pi$cm² **[1]**

7. $3^3 + \dfrac{1}{3}\pi \times 1.5^2 \times 4$ **[1]** $= 36.4$ **[1]** cm³ **[1]**

8. $\dfrac{4}{3}\pi \times 3.5^3 = a^3$ **[1]**

   $a^3 = 179.59$ **[1]**

   $a = 5.64$cm **[1]**

9. $\dfrac{2}{3}\pi \times 5^3$ **[1]** $= \dfrac{250}{3}\pi$ **[1]** cm³ **[1]**

Module 32: Pythagoras' Theorem and Trigonometry

1. **(a)** $\cos p = \dfrac{2.8}{3.5}$ **[1]** $= 0.8$ **[1]**

   $p = 36.9°$ **[1]**

   (b) $\tan 51° = \dfrac{KL}{10.5}$ **[1]**

   $KL = 12.966$ **[1]** $= 13.0$m **[1]**

2. $\dfrac{\sin A}{6.3} = \dfrac{\sin 63°}{8.1}$ **[1]**

   $\sin A = 0.693...$ so $A = 43.9°$ **[1]**

   Angle $ABC = 73.1°$ **[1]**

3. $9^2 = 7^2 + q^2$ **[1]**

   Substitute the values in the Pythagoras equation and then rearrange it.

   $q^2 = 9^2 - 7^2$ **[1]** $= 32$

   $q = \sqrt{32}$ (or $4\sqrt{2}$) **[1]**

4. **(a) (i)** $\dfrac{1}{\sqrt{2}}$ or $\dfrac{\sqrt{2}}{2}$ **(ii)** 1

   (b) $a = 2 \times \sin 60° = 2 \times \dfrac{\sqrt{3}}{2}$ **[1]** $= \sqrt{3}$cm **[1]**

   $b = 2 \times \cos 60° = 2 \times \dfrac{1}{2}$ **[1]** $= 1$cm **[1]**

5. $\sqrt{14^2 + 6^2 + 3^2}$ (or clear intention to find 3D diagonal) **[1]**
   $= 15.5$cm **[1]**

   The longest pencil will fit diagonally across the box.

Module 33: Vectors

1. $\mathbf{a} = 2\mathbf{r}$ $\mathbf{b} = -3\mathbf{s}$

   $\mathbf{c} = \mathbf{r} + \mathbf{t}$ $\mathbf{d} = 2\mathbf{t} - \mathbf{s} = \mathbf{r} + \mathbf{t} - 2\mathbf{s}$

   [1 for each correct answer]

2. **(a)** $\overrightarrow{BE} = -2\mathbf{b}$

   The opposite sides of a parallelogram are described with equal vectors.

   (b) $\overrightarrow{AC} = \mathbf{b} + \mathbf{d}$ **[2]**

   (c) $\overrightarrow{CE} = \overrightarrow{CB} + \overrightarrow{BE} = -2\mathbf{b} - \mathbf{d}$ **[2]**

3. $\overrightarrow{OR} = 2\mathbf{q}$ **[1]**

   $\overrightarrow{PQ} = \mathbf{q} - \mathbf{p}$ **[1]**

   $\overrightarrow{PR} = 2\mathbf{q} - \mathbf{p}$ **[1]**

4. If $\overrightarrow{OP} = 2\mathbf{p}$ and $\overrightarrow{OQ} = 2\mathbf{q}$ then $\overrightarrow{PQ} = 2\mathbf{q} - 2\mathbf{p}$ **[1]**

   And $\overrightarrow{OA} = \mathbf{p}$ and $\overrightarrow{OB} = \mathbf{q}$ giving $\overrightarrow{AB} = \mathbf{q} - \mathbf{p}$ **[1]**

   $\overrightarrow{PQ} = 2\overrightarrow{AB}$ therefore the lines are parallel **[1]**

   Write each vector in terms of the vectors that you are given and then look for a relationship.

Module 34: Experimental and Theoretical Probability

1. **(a)** $\dfrac{4}{14}$ or $\dfrac{2}{7}$ **(b)** $\dfrac{5}{14}$ **(c)** $\dfrac{9}{14}$

2. $\dfrac{1}{4} \times 120$ **[1]** $= 30$ **[1]**

   Each time the probability she will be correct is a quarter, since there are four different suits.

3. $1 - (0.2 + 0.3 + 0.25 + 0.1)$ **[1]** $= 0.15$ **[1]**

4. **(a)** $\dfrac{12}{25}$ or 0.48

   (b) $3 + 4 + 12 = 19$ **[1]**

   $\dfrac{19}{25} \times 250$ **[1]** $= 190$ **[1]**

5. $\dfrac{3}{4} \times \dfrac{1}{4}$ **[1]** $= \dfrac{3}{16}$ or 0.1875 **[1]**

   P(not 4) AND P(4) gives the probability of a 4 in exactly two spins.

Module 35: Representing Probability

1. **(a)**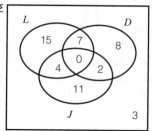

   [3 if fully correct; 2 for six correct numbers; 1 for three correct numbers]

   0 in the middle since no one can vote for all three.
   All the numbers in the diagram must add up to 50.

   (b) $\dfrac{11}{50}$ **(c)** $\dfrac{7}{26}$ **[1 for '7' and 1 for '26']**

2. **(a)**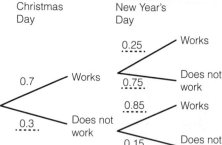

   0.3 **[1]**
   0.25, 0.75, 0.85, 0.15 **[1]**

   (b) P(Christmas AND not New Year) OR P(Not Christmas AND does New Year) =
   $(0.7 \times 0.75) + (0.3 \times 0.85)$ **[1]** $= 0.525 + 0.255 = 0.78$ **[1]**

   All four probabilities are different on the second branch because it is conditional on the first branch. Remember to multiply along the branches and then add up the two separate possible outcomes.

Module 36: Data and Averages

1. **(a)** 498   **(b)** 499

   The median is between 498 and 500 and so is 499.

   (c) Suitable answers, e.g. The sample size is too small. Need to sample across the range of times during production.
   [1 for each reason up to a maximum of 2]

2. **(a)** $60 \leqslant w < 80$

   Median value is between the 80th and 81st value, which is in the $60 \leqslant w < 80$ class interval.

   (b) Correct midpoints (35, 45, 55, 70, 90, 110, 135) **[1]**
   Correct total of midpoint × frequency of 10 915 **[1]**
   Estimated mean = 68.2g **[1]**

   (c) $20 + 25 + 22 + 44 = 111$ **[1]**
   $111 \div 160 = 69.4\%$ **[1]**

3. 6, 6, 6, 7, 10 or 6, 6, 6, 8, 9 **[3]** Other solutions are possible.
[1 mark for any one of and 2 marks for any two of: a total of 35, median of 6 or mode of 6]

> The total of all five numbers must be 35, i.e. 5 (cards) × 7 (mean) = 35

Module 37: Statistical Diagrams
1. (a) 21
(b) A bar chart or vertical line graph, with frequency on the y-axis and days of the week on the x-axis, e.g.

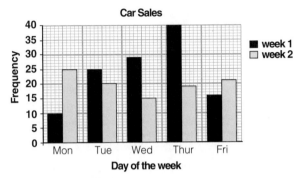

[1 for selecting a dual bar chart or vertical line diagram; 1 for fully correct data (frequencies for each bars); 1 for a fully labelled chart]

2. Table completed as follows: Seafood – 106.7°; Meat – 66.7°; Chicken – 53.3°; Mushroom – 55 and 73.3° **[1 for correct frequency of mushroom and 1 for correct angles]**

> Total frequency must be 45 × 6 = 270 as vegetarian represents one-sixth of the pie chart. 270 – 215 = 55 mushroom

3. (a) Attempt to draw a graph to interpret the data **[1]**
Correct labelling of axes **[1]**
Appropriate line of best fit drawn **[1]**
Estimate 175–195 million **[1]**
(b) Any suitable answer, e.g. It would not be reliable to use the graph to predict sales in year 14 because it is beyond the range of the data **[1]** and the trend of sales may change **[1]**.

4. (a) A scatter graph drawn with age on x-axis and speed on y-axis **[1]** and the points correctly plotted **[1]**. Data supports John's conclusion. **[1]**
(b) Negative correlation

> The trend shows that the younger the driver, the higher the speed; and the older the driver, the lower the speed. So the correlation is negative.

5. Attempt to find the frequencies using the histogram **[1]**
Process to use frequency and midpoints **[1]**
Method of total (frequency × midpoint) ÷ total frequency **[1]**
Mean of 36–37 **[1]**
Correct conclusion that the data supports an increase in travel distances **[1]**

Module 38: Comparing Distributions
1. (a) Class A: 48.8 **[1]**, Class B: 45.7 **[1]**
(b) Class A: Min 23, LQ 34, Median 45, UQ 62, Max 83 **[2]**
Class B: Min 27, LQ 32.5, Median 42, UQ 52.5, Max 78 **[2]**

> Order the numbers smallest to biggest. Then find the position of a quarter, half and three-quarters of the way through the data.

(c) Comparison of either interquartile range (IQR) or median
Class A: IQR = 62 – 34 = 32; Median 45 **[1]**
Class B: IQR = 52.5 – 32.5 = 20; Median 42 **[1]**
Any suitable conclusion, e.g. Class B is more consistent but has a lower median **[1]**.

2. (a) Cumulative frequency curve drawn correctly with points (150, 10); (200, 48); (250, 96); (300, 127); (350, 147); (400, 155); (500, 160)
[3 if fully correct; 2 if one error; 1 for correct cumulative frequencies]

> Remember cumulative frequency is a running total of the frequency column.

(b) Minimum = 100; consistent with the curve (values at 40, 80 and 120) for LQ = 190 (±5), median = 230 (±5), UQ = 285 (±5); maximum = 500 **[2 if fully correct; 1 if one error]**

> 160 people in total so LQ will be found at 40th, median at 80th and UQ at 120th. Draw lines across from the y-axis at these points and read off where they hit the curve from the x-axis.

(c) Any suitable answers, e.g. compare the IQR for both sets of data (approx. 100 and 160) so greater variation for younger drivers; compare the median for both sets of data (230 and 450) so much higher for younger drivers.
[1 for each comparison]
(d) Approx. 22 people **[1]**. This is an estimate as the data has been grouped and actual costs are unknown within a group **[1]**.

Exam Practice Paper 1
1. Mean
2. (a) –5 – 3 = –8 **[1]**
–8 ÷ 4 = –2 **[1]**
(b) $a = -1$ and $b = 7$, $a = -4$ and $b = 4$, $a = -11$ and $b = 2$, $a = -25$ and $b = 1$
[1 mark for each correct pair of answers]

> Since a must be a negative number, the solution $a = 1$, $b = 14$ is invalid.

3. Ann got two lots more than Timmy.

1 lot $\frac{£32}{2}$ = £16 **[1 for showing correct division]**
George got (16) × 5 **[1 for your answer × 5]** = £80 **[1]**

4. (a) $\left(\frac{20 \times 9}{4}\right) \times 3$ **[1]** = £135 **[1]**

> Round each number to 1 significant figure.

(b) 885 × 22 = 19 470 **[1]**
8.85 × 22 = £194.70 **[1]**

5. $8x - 5y = 19$
$60x + 5y = 15$ **[1]**
$68x = 34$ **[1]**
$x = \frac{1}{2}$ **[1]**
$y = -3$ **[1]**

6. P(orange) = $\frac{2}{3}$ **[1]**
$\frac{2}{3}$ of 9 = 6 **[1]**

> To start with there was a multiple of 3 chocolates and then a multiple of 4 after 1 had been eaten. So there must have been 9 chocolates to start with.

7. 10% = 34, so 5% = 17 **[1]**
340 + 34 + 17 = £391 **[1 for correct answer including units]**

8. (a) $\frac{4}{6}$ **[1]** $= \frac{2}{3}$
(b) $BC^2 + 4^2 = 6^2$ **[1 for correct use of Pythagoras]**
$BC = \sqrt{6^2 - 4^2} = \sqrt{20}$
$\tan B = \frac{4}{BC}$ **[1]** $= \frac{4}{\sqrt{20}}$ **[1]**
$= \frac{4\sqrt{20}}{\sqrt{20}\sqrt{20}} = \frac{\sqrt{20}}{5}$ **[1]** $= \frac{2\sqrt{5}}{5}$

9. $\frac{15}{16}$

> Turn the second fraction upside down and multiply.

10. $x^2 - 8x - 9 = 11$ [1]

$x^2 - 8x - 20 = 0$ [1]

$(x+2)(x-10) = 0$

$x = -2$ or $x = 10$ [2]

11. $81x^0y^{00}$ [2]

12. Using the bar chart to find correct total of years
    1, 2, 3, 5 = 95 [1]
    TV sales: 120 – 95 = 25 (i.e. 25 000) [1]
    Computer sales: Solving $2a + 5 = 25$, so $a = 10$ (i.e. 10 000) [1]

13. $m = 2$, $c = -3$ [1]

$y = 2x - 3$ [1]

Gradients of parallel lines are equal.

14. (a) Sides of triangle $x + 3$ [1]
    Perimeter of triangle = perimeter of square
    $3x + 6 = 4x$ [1]
    $x = 6$ [1]

    (b) Height of triangle = $\sqrt{9^2 - 3^2}$ [1 for correct use of Pythagoras on half of triangle] = $\sqrt{72}$ [1]

    Diagonal of square = $\sqrt{6^2 + 6^2} = \sqrt{72}$ [1] Therefore equal.

15.

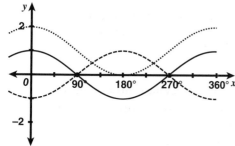

(a) See dashed curve above [1 for correct shape; 1 for correct position]

(b) See dotted curve above [1 for correct shape; 1 for correct position]

16. $\pi \times 3 \times 2$ [1] = 6πcm [1]

17. $\angle BAC = 50°$ [1] (alternate angles are equal)
    $\angle BCA = 180 - (50 + 75) = 55°$ (angle sum of a triangle = 180°)
    $\angle BCD = 180 - (55 + 80) = 45°$ (angles on a straight line = 180°)
    [1 for correct answer; 1 for three correct reasons]

18. Differences are 5, 11, 17
    So next difference is 23 [1]
    So next term is 54 [1]

    You need to go to the second difference to find the pattern.

19. $3x + 5 + x + 8x - 43$ [1]

$12x - 38 = 240 - 50$ [1]

$x = 19$ [1]

$\dfrac{19}{240}$ [1]

Write and solve an equation to find x.

20. After increase, let 1 part be x so values are $5x$ and $2x$
    [1 for a method using one ratio]
    After decrease, values are $5x - 25$ and $2x - 25$ and these are in ratio 5 : 1
    $5x - 25 = 5(2x - 25)$
    $100 = 5x$ and $x = 20$ [1 for correct calculation of a part]
    After increase values are $5 \times 20 = 100$ and $2 \times 20 = 40$
    [1 working back to original values]
    So original ratio 80 : 20 = 4 : 1 [1]

    This could be done by trials, which is quicker!
    5 : 2 …could be £50 : £20 but then original value of B = £0
    …could be £100 : £40 and original values £80 : £20 = 4 : 1
    and after £5 off £75 : £15 = 5 : 1

An alternative way of solving this question would be:
$(a + 20) : (b + 20) = 5 : 2$
Writing the ratio as a fraction gives: $\dfrac{(a+20)}{(b+20)} = \dfrac{5}{2}$
Cross-multiplying gives: $2a - 5b = 60$
Similarly for $(a - 5) : (b - 5) = 5 : 1$
Giving equation: $a - 5b = 20$
These two equations can be solved giving $a = 80$, $b = 20$
and ratio 4 : 1

21. (a) $4\sqrt{3}$

    Change it to $\sqrt{16} \times \sqrt{3}$

    (b) $\dfrac{2}{3+\sqrt{7}} \times \dfrac{3-\sqrt{7}}{3-\sqrt{7}}$ [1] = $\dfrac{6-2\sqrt{7}}{9-(\sqrt{7})^2} = \dfrac{6-2\sqrt{7}}{2} = 3 - \sqrt{7}$ [1]

22. 15% of 260 = 26 + 13 = 39 [1]
    18% of 210 = 21 + 21 – 4.2 = 37.8 [1 for any correct method]
    So 15% of 260 is greater [1 for stating the correct answer]

    In non-calculator percentage questions, there can be several ways to get the correct answer. Here 18% = 20% – 2% or 10% + 5% + 3%. Make sure your working is clear.

23. Any two suitable reasons, e.g. smaller spread so more consistent (interquartile range = 50 vs 155 hours) [1]; Higher median (= 370 vs 350) [1]

24. Scale factor for volume $\dfrac{810}{240} = \dfrac{27}{8}$ [1 – you could also write as a ratio 27 : 8]

    Scale factor for length = $\dfrac{\sqrt[3]{27}}{\sqrt[3]{8}} = \dfrac{3}{2}$ [1 for taking cube roots]

    Scale factor for area = $\dfrac{3^2}{2^2} = \dfrac{9}{4}$ [1 for squaring for area scale factor]

    Surface area of B = $180 \times \dfrac{9}{4}$ = 405cm² [1 for correct value]

25. $x^2 - 8x + 17 = (x - 4)^2 + 1$ [2]
    Turning point at (4, 1) [2]
    Crosses y-axis at (0, 17) [1]
    Line of symmetry $x = 4$ [1]

Exam Practice Paper 2

1. $425 \times 0.88 = £374$
   [2 for correct answer; 1 for any correct method]

   The multiplier method or working out 12% and subtracting are both correct. If you use a mental method on a calculator paper, you must get the answer correct to gain any marks.

2. 105 [1] and 210 [1]
   $420 = 2 \times 2 \times 3 \times 5 \times 7$ [1]
   $630 = 2 \times 3 \times 3 \times 5 \times 7$ [1]

   Use prime factor trees to find the two common factors
   $2 \times 3 \times 5 \times 7 = 210$ and $3 \times 5 \times 7 = 105$

3. $\pi \times 3.25^2$ [1]
   $\pi \times 3.25^2 \times \dfrac{3}{4}$ [1] = 24.887…
   24.9m² [1 for correct rounding and units]

4. (a)

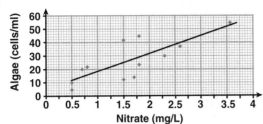

[2 if fully correct; 1 if one error]

(b) Weak positive

(c) Correctly read from line of best fit as shown above [1]: range 1.6–1.9mg/L [1]

(d) Predictions will only be reliable within the range of the nitrate levels in this data (i.e. 0.5–3.6mg/L).

5. (a) $4x^2 - 4x + 1$ [2]

(b) $8x^3 - 12x^2 + 6x - 1$ [2]

6. (a) $\dfrac{41.91607978}{7.529536}$ [1] = 5.566 887 49... [1]

> Work out the numerator and denominator separately and write them down.

(b) 5.57 (or your answer to part (a) correctly rounded)

7. $-2 = (-3)^2 + 2(-3) - 5$ [1]

$-2 = 9 - 6 - 5$

$-2 = -2$ [1]

> Substitute $x = -3$ and $y = -2$ into the equation. Remember negative integer rules.

8. $\dfrac{1}{3} \times \dfrac{1}{3}$ [1] = $\dfrac{1}{9}$ [1]

9. $60 : 100 = 3 : 5$ [1 for correct ratio]

$5.4 \times \dfrac{5}{3}$ [1] = 9m [1 for answer and units]

> You may use another correct method, e.g. 60cm × 5.4 = 900cm = 9m

10.

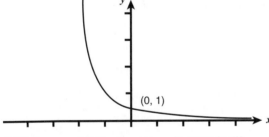

[2 if fully correct; 1 if curve passes through (0, 1)]

> Always ensure you show clearly where the curve crosses any axes.

11. (a) About 20°C

(b) Any answer in the range 22°C–25°C

(c) $\dfrac{43 - 20}{10} = 2.3$ or $\dfrac{20 - 43}{10} = -2.3$, cooling at rate of 2.3°/minute

[1 for using a tangent on the graph; 1 for correct calculation for your tangent; 1 for answer in range 2–3.3]

12. (a) First farm (no sale): 0.6 [1]

Second farm (from top to bottom): 0.4, 0.6, 0.4, 0.6 [1]

(b) $1 - (0.6 \times 0.6)$ [1] = $1 - 0.36 = 0.64$ [1]

> Calculate 1 – P(no sale, no sale)

13. $5y - xy = 10x - 2$ [1]

$5y + 2 = 10x + xy$ [1]

$5y + 2 = x(10 + y)$ [1]

$x = \dfrac{5y + 2}{10 + y}$ [1]

14. $\sin R = \dfrac{3.7}{5.8}$ [1] = 0.6379...

$\sin^{-1}\dfrac{3.7}{5.8} = 39.6377...$ [1] = 39.6° [1 for correct rounding and degree symbol]

15. (a) $\dfrac{27}{9}$ [1] = 3

$9 + 7 = 16$ [1 for adding parts]

$3 \times 16 = 48$kg [1]

(b) 10^6cm^3 in 1m^3 [1]

8900×1000g [1]

$8900 \times 1000 \div 1000\,000$ [1] = 8.9g/cm^3 [1]

(c) Density = $\dfrac{\text{mass}}{\text{volume}}$ so 9g of copper has a volume of

$\dfrac{9}{8.9}$ [1] = 1.01cm^3 and 7g of zinc has a volume of 1cm^3.

Total mass = 16g

Total volume = 1 + 1.01 = 2.01 [1 for mass and volume]

Density of alloy = 16 ÷ 2.01 = 7.96g/cm^3 (2 d.p.) [1]

16. $x^2 - 7x + 14 = 2x - 4$ [1]

$x^2 - 9x + 18 = 0$

$(x - 6)(x - 3) = 0$ [1]

$x = 6$ or $x = 3$

At $x = 6$, $y = 8$ [1] and at $x = 3$, $y = 2$ [1]

So coordinates are (6, 8) and (3, 2)

17. $x = 0.08\dot{1}\dot{}$

$10x = 0.8\dot{1}$

$1000x = 81.8\dot{1}$

$990x = 81$ [1]

$x = \dfrac{81}{990}$ [1] = $\dfrac{9}{110}$ [1]

> Label the recurring decimal as x and then multiply by 10 and 1000, so that when you subtract them the answer will be an integer, as the recurring decimals have cancelled each other out.

18. New gradient is $\dfrac{-1}{(-2)} = \dfrac{1}{2}$ [1]

$y = \dfrac{1}{2}x + c$ [1]

Substituting $x = 8$, $y = -1$ [1]

$-1 = 4 + c$

$c = -5$ [1]

$y = \dfrac{1}{2}x - 5$

19. (a) 1 : 9

(b) $10\,800 \div 27$ [1] = 400ml (or 400cm^3) [1]

20. $\dfrac{KM}{\sin 100°} = \dfrac{KL}{\sin 35°}$ [1 for correct application of sine rule]

$KL = \dfrac{10}{\sin 100°} \times \sin 35°$ [1] = 5.82cm [1]

21. $V_n = 4\big(5(n + 2) + 2\big) - 2(5n + 2)$ [1]

$V_n = 4(5n + 12) - 10n - 4$

$V_n = 10n + 44$ [2]

$V_n = an + b$ so is arithmetic sequence [1]

$U_n = V_n \Rightarrow 10n + 44 = 5n + 2$ [1]

No, since no positive integer solution for n [1]

Exam Practice Paper 3

1. (a) Lysette, as she has done the most coin flips.

(b) $\dfrac{28}{80} \times 500$ [1] = 175 [1] or $\dfrac{44}{120} \times 500$ [1] = 183 [1]

2. $3x + 4 > 19 \Rightarrow x > 5$ [1]

$2x - 1 < 13 \Rightarrow x < 7$ [1]

So $x = 6$ [1]

3. (a)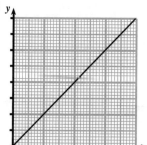

[1 for straight line with positive gradient; 1 for line from (0, 0)]

(b)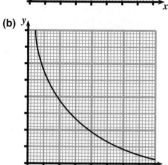

[1 for concave curve; 1 for lines approaching but not touching axes]

4. (a) [1 for correct use of scale]

Answer in range 420–450km **[1]**

(b) Using your answer from part (a) (for example we are using 440km) $\frac{700}{440} = 1.590 = 159\%$ Answer in range 155–165%

[1 for correct division; 1 for correct percentage]

(c) $\frac{700}{11} = 63.64 = 64$kmh^{-1} or 64km/h

[1 for method; 1 for correct answer; 1 for correct unit]

5. (a) 3^{-5} **(b)** 7.02×10^{-5} **(c)** 4

6. (a) $40 \leqslant h < 60$

(b) Midpoints (10, 30, 50, 70, 100) **[1]**

Sum of frequency × Midpoints = 7000 **[1]**

7000 ÷ 140 = 50cm **[1]**

(c) Correct calculation of $42 \div 140 = 30\%$ or $\frac{3}{10}$ **[1]**.

So the data does not support the farmer's claim. **[1]**

7. (a) 44 800km/h

(b) $(44\,800 \times 3) \div 2$ **[1]** = 67 200 **[1]** = 6.72×10^4km **[1]**

Remember to put your answer back into standard form, with any relevant units.

8. (a) [1 for constructing radii to B and D]

Angle $BOD = 360 - (90 + 90 + 52)$ **[1]** = 128°

Angle $BCD = \frac{1}{2}$ angle $BOD = 64°$ **[1]**

(b) $CBED$ is cyclic quadrilateral; angle $BED = 180 -$ angle BCD **[1]**

= 116° **[1]**

9. (a) £18 000

(b) $18\,000 \times 0.85 \times 0.85$ **[1]** = £13 005 **[1]**

(c) $18\,000 \times 0.85^4 = £9396.11$ **[1]**

4 years **[1]**

10. $6(2x+2) - 4x = 2x(2x+2)$ **[1]**

$8x + 12 = 4x^2 + 4x$ **[1]**

$x^2 - x - 3 = 0$ **[1]**

In a quadratic such as this, remember you can and should divide through by a common factor to make the numbers easier to deal with.

$x = \frac{1 \pm \sqrt{13}}{2}$

$x = 2.30$ **[1]** or $x = -1.30$ **[1]**

11.

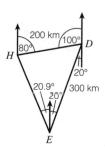

$180 - 100 - 20 = 60°$ **[1 for correct method to calculate angle at D]**

$EH^2 = 200^2 + 300^2 - 2 \times 200 \times 300 \times \cos 60°$ **[1 for correct substitution into cosine rule]**

$EH^2 = 70\,000$ so $EH = 264.575...$km **[1]**

$\frac{\sin E}{200} = \frac{\sin 60°}{EH}$

$\sin E = \frac{\sin 60°}{EH} \times 200$ **[1 for correct substitution into sine rule]**

$\sin E = 0.6546...$ $E = 40.9°$ **[1]**

Bearing = $360 - (40.9 - 20) = 339.1°$ or 339° **[1]**

12. Ratio of radii 4 : 5 so whole cone has height 25cm

[1 for attempt to find height using ratio or scale factors; 1 for finding height]

Volume of frustum = $\frac{1}{3}\pi \times 15^2 \times 25 - \frac{1}{3}\pi \times 12^2 \times 20$ **[1]**

= 2874.55 **[1]** = 2870cm^3 **[1 for correct rounding and units]**

13. $4x^2 - 24x + 41 = 4\left[x^2 - 6x + \frac{41}{4}\right]$ **[1]**

$= 4\left[(x-3)^2 - 9 + \frac{41}{4}\right]$ **[1]**

$= 4\left[(x-3)^2 + \frac{5}{4}\right]$ **[1]**

$= 4(x-3)^2 + 5$ **[1]**

$(a = 4, b = 3, c = 5)$

14. (0.7×0.7) **[1]** $+ (0.3 \times 0.8)$ **[1]** = 0.49 + 0.24 = 0.73 **[1]**

Draw a tree diagram to help visualise the probabilities.

15. Lower bound for bottle = 747.5ml. Upper bound for cup = 5.5ml **[1]**

$\frac{747.5}{5.5}$ **[1]** = 135.909... = 135 full cups **[1]**

16. (a) $g(3) = 7$ **[1]** $f(7) = 20$ **[1]**

In questions like these, it is often easier to substitute numbers in first, rather than work out an algebraic expression for $fg(x)$

(b) $f^{-1}(x) = \frac{1+x}{3}$ **[1]**

$g^{-1}(x) = \frac{x-1}{2}$ **[1]**

$\left(\frac{1+x}{3}\right)\left(\frac{x-1}{2}\right) = 4$

$x^2 - 1 = 24$ **[1]**

$x = \pm 5$ **[2]**

17. $(16^{\frac{1}{4}})^3 = 2^3$ **[1]** = 8 **[1]**

18.

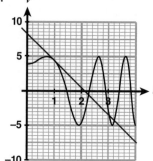

Tangent line drawn **[2]**

Gradient approximately $\frac{-8.7}{2.1}$ **[2]**

Acceleration is approx. -4.1m/s^2 **[1]**

When estimating the gradient here, it is easier (and more accurate) if you use the points where the tangent line crosses the axes.